T0269150

CAMBRIDGE LIBRARY COLLECTION

Books of enduring scholarly value

History of Medicine

It is sobering to realise that as recently as the year in which On the Origin of Species was published, learned opinion was that diseases such as typhus and cholera were spread by a 'miasma', and suggestions that doctors should wash their hands before examining patients were greeted with mockery by the profession. The Cambridge Library Collection reissues milestone publications in the history of Western medicine as well as studies of other medical traditions. Its coverage ranges from Galen on anatomical procedures to Florence Nightingale's common-sense advice to nurses, and includes early research into genetics and mental health, colonial reports on tropical diseases, documents on public health and military medicine, and publications on spa culture and medicinal plants.

The Vaccine Contest

When English surgeon William Blair (1766–1822) embarked on his career, he became familiar with the devastation caused by smallpox in urban areas. The virus was lethal to more than a fifth of the people infected, and the rest were at risk of long–term complications. The first effective vaccine against the disease had been developed by Edward Jenner, who had been made aware that smallpox infection was uncommon among milkmaids who had been exposed to a milder form of pox contracted from cows. Although Jenner's vaccine was made available soon after its public announcement in 1798, the objections by various sceptics deterred many from embracing the procedure. In this 1806 pamphlet, Blair employs the format of a dialogue between an anxious parent and an ardent vaccination opponent to convince Londoners of the benefits offered by the new vaccine. His account is complemented by a report from the Royal Jennerian Society.

Cambridge University Press has long been a pioneer in the reissuing of out-of-print titles from its own backlist, producing digital reprints of books that are still sought after by scholars and students but could not be reprinted economically using traditional technology. The Cambridge Library Collection extends this activity to a wider range of books which are still of importance to researchers and professionals, either for the source material they contain, or as landmarks in the history of their academic discipline.

Drawing from the world-renowned collections in the Cambridge University Library and other partner libraries, and guided by the advice of experts in each subject area, Cambridge University Press is using state-of-the-art scanning machines in its own Printing House to capture the content of each book selected for inclusion. The files are processed to give a consistently clear, crisp image, and the books finished to the high quality standard for which the Press is recognised around the world. The latest print-on-demand technology ensures that the books will remain available indefinitely, and that orders for single or multiple copies can quickly be supplied.

The Cambridge Library Collection brings back to life books of enduring scholarly value (including out-of-copyright works originally issued by other publishers) across a wide range of disciplines in the humanities and social sciences and in science and technology.

The Vaccine Contest

Being an Exact Outline of the Arguments ...
Respecting Cow-Pox Inoculation

WILLIAM BLAIR

CAMBRIDGE
UNIVERSITY PRESS

CAMBRIDGE
UNIVERSITY PRESS

University Printing House, Cambridge, CB2 8BS, United Kingdom

Cambridge University Press is part of the University of Cambridge.
It furthers the University's mission by disseminating knowledge in the pursuit of
education, learning and research at the highest international levels of excellence.

www.cambridge.org
Information on this title: www.cambridge.org/9781108078023

This edition first published 1806
This digitally printed version 2017

ISBN 978-1-108-07802-3 Paperback

THE

VACCINE CONTEST;

OR,

" MILD HUMANITY,

" REASON, RELIGION, AND TRUTH,

" AGAINST

" *Fierce, unfeeling Ferocity, overbearing Insolence, mortified*
" *Pride, false Faith, and Desperation.*"

BY WILLIAM BLAIR.

[Price Half a Crown.]

Printed by S. GOSNELL,
Little Queen Street, Holborn.

THE
VACCINE CONTEST:

OR,

" MILD HUMANITY,

" REASON, RELIGION, AND TRUTH,

" AGAINST

"Fierce, unfeeling Ferocity, overbearing Insolence, mortified
" Pride, false Faith, and Desperation;"

BEING

An Exact Outline

OF THE

ARGUMENTS AND INTERESTING FACTS,

ADDUCED BY THE

PRINCIPAL COMBATANTS ON BOTH SIDES,

RESPECTING

COW-POX INOCULATION;

INCLUDING A LATE

OFFICIAL REPORT ON THIS SUBJECT,

BY THE

MEDICAL COUNCIL

OF THE

ROYAL JENNERIAN SOCIETY.

CHIEFLY DESIGNED, FOR
THE USE OF CLERGYMEN, HEADS OF FAMILIES, GUARDIANS, OVERSEERS
OF THE POOR, AND OTHER UNPROFESSIONAL READERS WHO MAY
BE CONCERNED FOR THE WELFARE OF MANKIND.

BY WILLIAM BLAIR, M. A.

Surgeon of the Lock Hospital and Asylum, the Bloomsbury Dispensary, and New Rupture
Society; Member of the Royal College of Surgeons, and of the Medical
Societies of London, Paris, Brussels, Aberdeen, &c. &c. &c.

" When men have become so abandoned as to *pervert truth*, it is high time
" for the learned in the Faculty to awaken from their torpid lethargy."

DR. ROWLEY.

LONDON:
PRINTED FOR J. MURRAY, NO. 32, FLEET STREET.
1806.

PREFACE.

ALTHOUGH the Author has by no means been a supine or inattentive observer of the progress of vaccine inoculation, but has very faithfully endeavoured to appreciate the facts and arguments on both sides, as so important a question required, he is not one of those who make the new practice a leading branch of his profession. He believes, however, that he was *among the first* to introduce and encourage the Cow-pox, without the impediment of exacting the usual recommendatory letters, at a public Infirmary; from whence, likewise, he has cheerfully and *gratuitously* distributed, a considerable number of charges of vaccine fluid, both in town and country. The following printed form of invitation was ordered to be circulated by the Governors of the Bloomsbury Dispensary, at the request of their medical officers; in consequence of which the Author has had an opportunity of inoculating above six hundred persons, exclusive of his private patients:

" *Bloomsbury Dispensary, No.* 62, *Great Russel Street*; *May* 15, 1802.

" We hereby give notice, That all poor persons
" desirous of being inoculated for the Cow-pox

" (without any expense or letter of recommendation),
" may attend the Dispensary for that purpose, on
" Tuesdays, at half-past twelve o'clock. And we
" recommend to those who regard the health of
" their children, or the safety of their friends and
" neighbours, to avail themselves of the opportunity
" now offered: as experience has shewn, in many
" thousands of instances, that the Cow-pox is com-
" pletely effectual in preventing the Small-pox ; that
" it is so mild, as to be never attended with any ha-
" zard, or even a personal blemish ; and is not com-
" municable by the breath or perspiration, but by
" inoculation only.

    " J. CARMICHAEL SMYTH, Consulting Physician.
    " GEORGE PINCKARD, Attending Physician.
    " EDWARD JENNER, M. D. Superintendant of the
        Vaccine Inoculation.
    " WILLIAM BLAIR, Surgeon."

During the preceding year, it was too manifest
that prejudices and apprehensions were assiduously
stirred up, by the different objectors to vaccine practice ; some of whose names and deeds will probably
be long remembered by the public, with feelings of
honest indignation. The influence of false rumours
and distorted facts operated so strongly in the district
of Bloomsbury and St. Giles, as to preclude even a
single person from applying for inoculation at that
Dispensary, after the 12th day of November last,

until the present time! But it was not before the beginning of March 1806, that the Author thought it his duty to read a certain vaunting treatise of one of these active objectors; whose puffing hand-bills, news-paper paragraphs, and degrading *placards* upon the dead walls of London, had previously informed him of its tendency.

Truth will always bear looking in the face. It is never ashamed to be closely examined, by the most scrupulous and penetrating eye: but Truth itself may be so varnished over, and perverted by ingenious glosses, as to appear unlike herself, and even to be mistaken for Error. Such, then, seems to have been the unhappy effect of Dr. Rowley's artful publication on the Cow-pox; than which, the author of these pages has never perused a literary work, on any subject, so completely unfair—so insidiously imposing—so calculated to deceive—so mischievous in its tendency—so full of invective—so abounding in falsehoods—so plainly inconsistent with itself—and so disgustingly repugnant to common decency.

A mere glance at this book awakened attention; and a deliberate perusal of it, suggested the idea of turning against an implacable adversary the murderous weapons which he himself had provided for a different purpose. The Author judged it would not be lost time, though a nauseous and revolting task, to extract the *marrow* or *quintessence* of that extraordinary performance; and, by placing THE DOCTOR'S OWN LANGUAGE in a new and vivid light, to afford a spirited and glowing picture of its genuine deformities. The real character and motives of an opponent who is so entirely devoid of justice and decorum, cannot be better discovered, than by dissecting, analysing, and exposing to public view, what may be called the vitals and sinews—the internal springs—the peculiar features and tone, of his composition. And, if it should be found, that his character and motives are far from pure, except in his own eyes, it may be questioned whether his pretended TRUTHS be *unexceptionable,* or his *alledged* FACTS such as honesty demands.

In attempting this ungrateful task, (lest the Author might be supposed to act from personal pique or resentment, on account of insults received, in common with many of his medical friends,) it may here be proper to suggest, that he is *one of the few* practitioners whom Dr. Rowley has never offended ; but, on the contrary, who has chosen to compliment him in print, for " his INGENUITY, CANDOUR, and LOVE OF TRUTH." If, in performing the present duty, he should in reality appear to deserve the commendation he has unexpectedly received, from a writer whose praise is worse than blame, it will yield him an abundant recompense for the irksomeness of his labour.

The portion of materials extracted from Dr. Rowley's work, being in detached sentences and clauses, made it necessary to give the whole a *colloquial form* ; by which means the Author was enabled to introduce all the striking paragraphs and phrases, while at the same time he has preserved a general consistency and uniformity of style. Perhaps, likewise, the singularity and boldness of sentiment which are thus infused, will excite the greater interest, and produce more feeling in the unprofessional reader than could otherwise have been looked for, on a medical subject.

———————

P. S. While this Pamphlet was in the press, the Author heard that Dr. Rowley had been summoned to another, and more awful, tribunal than that of *Man !* As the following pages, in defence of TRUTH, are of equal concern to the public, whether the great Champion of anti-vaccination be living or not ; and, as the errors he propagated so industriously are still kept alive by others, and are unlikely to be soon eradicated ; the Author could not think it his duty to refrain from publishing what he had conscientiously prepared for the good of mankind. He was never in the smallest degree prompted by personal motives, to take up his pen in this important cause of humanity ; and therefore could see no reason, on the death of an individual, to relinquish the design in which he had engaged. The foregoing prefatory observations are precisely what remained in the Printer's hands, when the event alluded to was first communicated to the Author, on the 18th March 1806.

Great Russel Street,   W. BLAIR.
  Bloomsbury.

THE
VACCINE CONTEST:

OR,

" MILD HUMANITY,
" REASON, RELIGION, AND TRUTH,

" AGAINST

" *Fierce, unfeeling Ferocity, overbearing Insolence,*
" *mortified Pride, false Faith, and Desperation.*"

IT may be right to premise, that the curious TITLE to the sub-joined Dialogue is borrowed from the Hero of anti-vaccination. A Clergyman, residing in a village near London, is supposed to be deeply distressed and apprehensive, lest his remaining children should die of an epidemic small-pox; he, therefore, comes to Saville Row, and consults Dr. Bragwell on the steps necessary to be taken : but, before the conclusion of his discourse, a town-Surgeon enters the room, and joins in the conversation; by which means, the Rev. Gentleman, having previously been dissatisfied with Dr. Bragwell's indecorous behaviour, is fully convinced that " mild Humanity, Reason, Religion, and Truth," unite in favour of Vaccination.

PARENT.—Most renowned and learned Doctor Bragwell: I am the curate of a populous parish, and the afflicted parent of a numerous family. My first

darling babe was snatched away by the small-pox,
after Suttonian inoculation.   I have lost two others,
by the natural disease ; and am now in danger of being unhappily bereaved of all the rest (just returned
from school) by an epidemic small-pox, which rages
fatally in our neighbourhood, especially among persons who have cried down Vaccination.   Having heard,
Sir, of your great exertions and disinterestedness in
ascertaining the truth, amidst " the continual hurry
" of a most extensive practice," I trust you will for-
give my anxiety and freedom, in thus craving your
opinion on an important question, viz.—Whether I
should follow the advice of my honest Apothecary,
who recommends me to let him immediately vaccinate the dear children, and who has humanely offered
to inoculate for the Cow-pox *gratuitously* through
our whole village.

DOCTOR.—" Greasy horse-healed project ! The
" sooner Cow-pox infatuation is abandoned *in toto*,
" so much the better for society.   The credulous Fa-
" culty have been imposed on, until TRUTH hath
" opened their eyes to conviction."

PARENT.—Good Sir, how fortunate that I consulted you !   I thankfully acknowledge this providential interference.   I only wish to know the real
truth ; and should be ever obliged to you to explain
wherein " the Faculty have been imposed on."   But,
give me leave, Doctor Bragwell, to add, that our
Royal Princes, Noblemen, Bishops, " Clergymen,
" Ladies of Rank, and many others, the warmest

" and well-intentioned friends of humanity and phi-
" lanthropy, have united with great zeal, vehe-
" mence, and alacrity, to promulgate, recommend,
" and extend the utility of this novel Cow-pox prac-
" tice." Nay more, it is said that the Small-pox is
quite banished from particular towns, by this
simple means : and I am informed, on the best authority, that the Rev. Rowland Hill has vaccinated
nearly five thousand persons with his own hand, and
directed the vaccination of several thousands more ;
without so much as one unfavourable case, or
failure ! ! This particular anecdote I have learnt in
the stage-coach, as I came to town, from an intimate
acquaintance of Mr. Hill ; and it cannot, therefore, be called in question. But still, learned Sir,
I am all attention, to hear what facts and rea-
sons there are against this new practice, in other
quarters.

DOCTOR.—As for the great folks, I answer,
that " puffing letters, and letters from humane and
" benevolent Lords, prove nothing. When men are
" so uninformed, or have become so abandoned as
" to pervert truth, it is high time for the learned in
" the Faculty to awaken from their torpid lethargy.
" Those will be considered the greatest enemies to
" society, who longest persist in spreading the cri-
" minal and murderous evil."

PARENT.—With great deference to your superior judgment, let me ask, if those Noble Lords,

Ladies, and Reverend Gentlemen, may not bear witness to what they have seen with their own eyes, perhaps in their own families and neighbourhood, and wherein they cannot easily be deceived? I do not here presume to contradict *you*, Doctor Bragwell, who, doubtless, understand the subject perfectly; but I merely desire to learn, how the affirmation of such respectable and impartial eye-witnesses can be got rid of? If there were no friends to this practice but medical men (I beg pardon for the suggestion), some uncharitable and inconsiderate people might surmise, that their testimony is rather suspicious, as they may possibly be interested; though I myself see clearly, that their pecuniary interest lies much the other way, in giving all possible encouragement to so dreadful and loathsome a disorder as the small-pox.

DOCTOR.—I perceive, Reverend Sir, that you are more inquisitive, and less easily persuaded, than my patients and hearers in general. Let me, in turn, ask you a question; " Who are those vaccinators? " Have they distinguished themselves by deep me- " dical erudition," as I have? " Novices have dared " to cavil with veterans in science, with those who " had long experience before any of the furious vac- " cinators were born! Infatuated visionists! Daring " projectors, who bid defiance to truth! It is just to " infer, that millions will become victims to small- " pox, and perish by its epidemic malignity, who " are supposed secure through the imaginary infal-

" libility of Cow-pox inoculation. Dreadful in-
" stances of Cow-pox failure, &c. are pouring in from
" all quarters."

PARENT.—Respected and honoured Doctor: you
really astonish and quite petrify my inmost soul! I
know not what to say; except that I came here desirous of learning the whole truth, good or bad.
Pray, Sir, for mercy sake, tell me all you know; and
do not let me return home, without being fur-
nished with the most unanswerable facts and arguments.

DOCTOR.—" Irrational and destructive prac-
" tice! Wild, light-headed adherents, who have
" distinguished themselves for ignorance!" Their
cause is " supported by subterfuge, evasion, and cri-
" minal proceedings."

PARENT.—Oh! Sir, God forbid. You harrow
up my feelings! Proceed, Sir, I entreat, and tell me
every jot and tittle; though you do not yet detail
any precise facts, that I might lay hold of, for the
conviction of such of my parishioners as are carried
away by error and prejudice.

DOCTOR.—" It is a most glaring tyranny, and
" inhuman in the extreme, to force vaccination on
" the poor, *nolens volens*, in most parishes through-
" out the united kingdom; but MURDER WILL COME
" OUT!"

PARENT.—Nothing can palliate such conduct,
if it be true, but the fullest persuasion that the Cow-
pox is a beneficial discovery of universal importance;

as, indeed, I had been hitherto disposed to imagine
it was.   But, go on, Dr. Bragwell.

DOCTOR.—" The world did not require Cow-
" pox ; it was forced, contrary to inclination, into
" it.   The Cow-pox was forced into the world, with
" the utmost vehemence.   It was too hot to hold :
" therefore the refined artists struck briskly, while
" the iron was hot.   Many men of the strongest
" passions were first seized with this Cow-pox
" *mania*.   Some were so tyrannical and insolent, as
" to force parents to comply with the project."

PARENT.—How zealous ! How active, in ex-
tending vaccination ! I feel these remarks keenly, Sir:
I know well what anxiety tender parents must experience ; but you will presently explain yourself more
fully on this delicate point.

DOCTOR.—" They exceeded in zeal even the
" chimerical *Illuminati*, Swedenborg and his en-
" thusiastic *raving-mad* followers ! Despotic, in-
" human tyranny !   All was violence, blaze, uproar,
" and tumult ! Parliament sanctioned the experi-
" ment ; and parents, affectionate, unsuspicious pa-
" rents, were cajoled to receive beastly diseases, and
" to sacrifice their innocent infants, whether they
" would or not !"

PARENT.—Did not any honest man exclaim
against such proceedings ?

DOCTOR.—" Reason was trampled on ! Truth,
" *mild Truth*, stood astonished and silent, during the
" violence of this furious storm, and hid her virtuous

" head. Earth trembled ! and heaven profusely shed
" tears !"

" I have been in some vaccination storms ; and
" have had the buttons of my coat torn off, cloth and
" all, to convince me of the infallible excellence of
" cow-pox. I have seen some vaccinators redden
" like a flame, with fury ; the lips quivering, the
" eyes starting out of the head, with flashing streams
" of fire; the mouth foaming, and tongue pouring
" forth a torrent of hard words, like a thunder-
" storm ; the fist clinched like a pugillist, ready to
" accompany the violent wrath, with other knock-
" down arguments."

PARENT.—And how did you bear all this ?

DOCTOR. — " Mild investigating philosophy
" quits the scene, and leaves the field of battle to the
" BEDLAMITES. Virtue, inestimable virtue, is its own
" reward !"

PARENT.—You did right in not rendering evil
for evil, and railing for railing. Truth needs not
the aid of such rancour or bitterness.

DOCTOR.—" Truth is mild and gentle : false-
" hood is furious, revengeful, abusive, and malicious.
" Vaccine fury must cool, before some vaccinators
" return to justice, truth, or reason. Mild huma-
" nity, reason, religion, and truth, meet now in
" combat, against fierce, unfeeling ferocity, over-
" bearing insolence, mortified pride, false faith, and
" desperation. Fury, rage, and despair, are com-
" monly united to the worst causes. The most ex-

c

" cellent physicians ARE ALWAYS MODEST, CANDID,
" and UNASSUMING."

PARENT.—I much approve your just description
of an upright medical character, as well as your account of truth and falsehood. It is a pity there
should be any personal abuse or illiberal invective on
either side; but truth will finally reign.

DOCTOR.—" Facts are stubborn things, and
" must ultimately prevail. It is necessary always to
" know whether a man be a matter-of-fact-man, a
" strict adherer to NOTHING BUT TRUTH."

PARENT.—Most certainly; for if we once convict a person of habitually concealing, exaggerating,
or distorting facts, we must necessarily reject what
he says, however gravely or peremptorily he affirms
a thing to be true.

DOCTOR.—" Truth, absolute truth, never re-
" quires the props of low cunning, sophistry, or
" deception."

PARENT.—I should be extremely sorry to hear
that professional gentlemen, embarked in so good a
cause, condescended to use these unlawful weapons,
for defending the truth.

DOCTOR.—" A delicacy of sentiment, a regard
" for the honour of the art, and the characters of
" the erroneous, should ever prevent personalities.
" The greatest respect is due from one regular
" practitioner to another;" because " they all rank
" as gentlemen in this country, and too much ho-

" nour cannot be paid to so liberal and learned a pro-
" fession."

PARENT.—Excellent sentiments ! I admire this
liberality and politeness, worthy Doctor.

DOCTOR.—" Truth is the object of my inquiry;
" and to sacred Truth alone should contending par-
" ties make their appeal, and submissively abide by
" her decision. It has been my uniform conduct,
" through life, to detect fallacy, and to enforce *evi-*
" *dent truths*. When a boy, in 1757, I never suf-
" fered any hypothesis, or system, or idle specula-
" tion, to enter my mind ; and all my writings and
" lectures tend to these important purposes, how-
" ever they may have offended those who hate truth.
" It is lamentable to see a concourse of sagacious
" vaccinators sitting in partial judgment, raising
" questions, perplexities, and doubts, and making
" inquiries into abstruse hidden points ; and at the
" same time, daringly controverting and evading the
" most decided and *evident truths* militating against
" vaccination. What has truth and demonstration
" to fear ? Nothing. If vaccination be not trem-
" blingly afraid of truth ; if it stood on the firm basis
" of truth, it would not cavil, controvert, nor wax
" warm."

PARENT. Pray, learned Sir, how would you maintain the truth, against *opposers*, without " contro-
" verting" what they might advance ? But, I should
gladly hear from you what are " the most decided
" and evident truths" you speak of, " militating

c 2

" against vaccination." Having never met with your candid account of this vaccine practice, which the public newspapers and posting-bills announce, I beg the favour of you to resolve my doubts; lest, on parting from you to-day, I should follow the advice given me, to vaccinate my dear children without delay: for delay is highly dangerous, in their critical situation.

DOCTOR.—Although "I have no time for con-
" troversy," I wrote my last book "as a small tri-
" bute of gratitude for the unbounded confidence
" I have enjoyed among all ranks, from the highest
" to the lowest, during the whole period of my
" professional life; amidst the most active scenes in
" medicine, which perhaps ever fell to the lot of
" one man. The rapid sale and universal ap-
" probation with which the two former editions of
" my work have been received, induced me to pro-
" duce a third edition, from a clear conviction that
" all mankind may benefit by attending to the truths
" and reasonings therein promulgated. To which
" are added my modes of treating the beastly
" new diseases produced from Cow-pox, explained
" by two coloured copper-plate engravings; and
" above five hundred dreadful cases of small-pox
" after vaccination, or cow-pox mange, cow-pox
" ulcers, cow-pox evil, or abscess, cow-pox morti-
" fication, &c. with my certain, experienced, and
" successful mode of inoculating for the small-pox,
" which now becomes necessary from cow-pox
" failure, &c."

PARENT—Cow-pox failures! Why, I thought there were royal establishments and committees to encourage vaccination, and that the modest Dr. Jenner had been rewarded with ten thousand pounds sterling for this infallible preventive of the small-pox!

DOCTOR.—" When we see bodies of men (in-
" sane as the credulous alchymists) advertising as
" committees, to enforce the necessity of vaccina-
" tion, after KNOWING so many instances of failure,
" after KNOWING so many proofs of various disastrous
" consequences, it is natural to inquire of what ma-
" terials these committees are composed ?"

PARENT.—Is it possible that those medical
" committees KNEW of many instances of failure,
" many proofs of disastrous consequences ?" What
failures ? What consequences ? Pray, Sir, keep
me no longer in painful suspense.

DOCTOR.—" It is natural to inquire of what
" materials these committees are composed ? Where
" they were educated, and how long they have
" studied and practised medicine ? Whether they
" have already distinguished themselves for learning,
" skill, and judgment in the profession, either by
" writing or successful practice ; or, whether they
" be some of those versatile characters, issuing from
" those hot-beds of hypothesis, some of the plau-
" sible medical schools, who (having no fixed prin-
" ciples to guide them, originating from real and
" actual bedside practical experience and observa-

" tion) are always most prompt and foremost in
" retailing every new whim or imaginary conceit
" that makes its appearance? Whether they be
" men of sound intellects, and who are strictly
" guided by truth; or, whether they be men of
" warm imaginations, fond of noisy altercation, and
" appearing conspicuous, more as overbearing wran-
" glers than as men of solid judgment and lovers of
" truth? Whether they know the right use of
" reason or logic, in searching after truth; or, whe-
" ther they take up false propositions for true?
" These, and many such-like questions, strike sober,
" sedate, and reflecting minds, who are not seduced
" to swerve from truth."

PARENT.—Stop, for one moment, Doctor Bragwell. I am really bewildered by your multiplicity of
queries; and cannot think that any qualifications are
wanting on the present occasion, besides the faculty
of seeing, with plain common sense, and a moderate
share of experience in this department of physic. I am
unwilling to imagine you take the majority of those
practitioners for downright fools and rogues who
have given their decided sanction to Dr. JENNER's
plan of inoculating; and yet I find it difficult to put
any better construction upon your words!

DOCTOR.—I have distinctly told you, that these
" committees enforced vaccination after *knowing* so
" many instances of failure, &c."

PARENT.—Oh, Sir! how uncharitable! What
has become of your candour, your liberality, your

esteem for professional character, your regard for the honour of the art, and of regular practitioners? Is it possible? So many admired and learned physicians, so many eminent surgeons, &c.—all liars, cheats, knaves? Have they indeed asserted and published what they "KNOW" to be absolutely false? This is not to be hastily credited.

Besides, Sir, I remark that you are talking of " logic," and " reason," and " propositions," and " wranglers ;" as if vaccination had been a matter of dispute at " the university of Oxford ;" when you studied physic there about fifty years ago, and began to write your truly " elaborate work, *Schola Medicinæ* " *Universalis Nova*, in the original Greek and Latin :" whereas, I took vaccine inoculation for a recent practice, which you never before heard of, and therefore cannot have learnt by your academical education !

Although I don't pretend to understand the nature and effects of the Cow-pox, like you, Doctor Bragwell, I can tell when you are throwing dust into my eyes, that I may not distinguish truth from error. I beg pardon, if, in my warmth, this language be too plain ; but I sincerely desire to come at the truth, the whole truth, and " nothing but the truth," in so momentous an affair. It is not for me to trifle or hesitate, on a question which involves the life of my children.

DOCTOR.—" There are many suffering thou- " sands in the different districts of London, and in " most parts of the country, who execrate and curse

" the invention. Though poor, they do not think
" the lives or health of infant human beings are to
" be sported with, to gratify the chimerical specu-
" lations of dangerous curiosity. Are affectionate
" parents, during this bloody contest, to be robbed
" of the security which small-pox inoculation af-
" fords? If they any longer justify cow-poxing, they
" merit all they are doomed to suffer.—Let man-
" kind arouse from their vaccinating lethargy, and
" chase from their houses all who propose vacci-
" nation. Who would marry into any family, at
" the risk of their offspring having an hereditary,
" filthy, beastly disease? Are men to become the
" victims of horrid, beastly, chronic diseases ; vic-
" tims diseased for life; and transmit them to pos-
" terity for ages ; that a few fanatics in science MAY
" REVEL IN WEALTH? Forbid it heaven; forbid it
" humanity ; forbid it reason, justice, and truth!
" It is enough to freeze the soul with horror!"

PARENT.—Oh, Doctor, pray, Doctor Bragwell,
don't fly into such a violent rage. I thought " *Truth*
" was mild and gentle ;" but, alas! alas! " she hides
" her virtuous head."

DOCTOR.—" The Cow-POXERS listen to nothing.
" Let not their fruitless irascibility vent itself in
" gross scurrility and abuse ; for vulgar abuse and
" low sneers, prove nothing, nor will they invali-
" date SELF-EVIDENT FACTS. The five hundred
" and four facts against the continuance of cow-pox
" inoculation, out of which seventy-five have died.

" devoted victims to the filthy project, are but as a
" drop of water added to the sea ! There's scarcely
" a street, lane, court, or alley, through London,
" which does not afford some disastrous circum-
" stances arising from Cow-pox inoculation. In all
" parts of England, and on the Continent, may be
" found innumerable similar disasters. Whoever
" reads the public papers, will perceive advertise-
" ments of the venders of VELNO's *Vegetable Syrup,*
" declaring the numerous applications for that cele-
" brated remedy to remove the horrid effects of
" Cow-pox inoculation."

PARENT.—Passing strange ! Are you, then, so
perversely incredulous, when regular and authorized
practitioners speak ; but so completely credulous,
when irregular and unauthorized empirics puff off
their quack medicines in the " public papers ?". If
your " visible, indubitable, and self-evident facts"
are built on such a sandy foundation as. this, Doctor
Bragwell ; I must acknowledge the world has equal
reason to believe the pretensions of Dr. Rowley,
Dr. Squirrel, Dr. Brodum, Dr. Moseley, or " the
" venders of Velno's celebrated remedy." If you expect me, Sir, to believe your bold assertion, that the
five hundred and four unfavourable cases which you
have so diligently collected " ARE BUT AS A DROP OF
" WATER" compared " TO THE SEA," I must demand some better proof than your mere word, or the
ipse dixit of your advertising fraternity.

DOCTOR.—" There is SCARCELY A WEEK passes
" that I do not prescribe for some miserable case

" or other. I am prescribing EVERY DAY for all
" those horrid complaints which have followed Cow-
" poxing. The parents and friends in general are
" so exasperated against Cow-pox, that they seldom
" apply to the vaccinators."

PARENT.—This is, I think, an additional reason
for not crediting your tale of woe to its full extent ; as it may be supposed, under such favourable
circumstances, that you not only would amply repay
yourself for all this disinterested trouble, but also
might have obtained a clear and particular account of
those other " *suffering thousands*" in every street,
lane, court, and alley through London.

DOCTOR.—" I am ready to meet publicly the
" whole host of vaccinators, or their protectors, TO
" DEMONSTRABLY PROVE ALL MY ALLEGATIONS
" TRUE ,"—but " *humanum est errare*," and " any
" errors in the statement of facts will be readily
" corrected."

" The vaccinators, besides deluding parents, tam-
" pering with, or bribing the unfortunate victims to
" vaccination, threaten them unmercifully with
" ruin, if they dare to disclose vaccination errors.
" Others are cajoled, and forced to say or do any
" thing vaccination fallacy chooses to dictate. They
" endeavoured to crush to atoms every *cool* ob-
" server and reasoner, who dared to examine their
" doctrines. Thus the most sacred truths were per-
" verted by artifice and low stratagem. Mankind
" have been so repeatedly deceived, that if vacci-

" nation even spoke truth, it would not be believed
" by any one who has the least pretension to dis-
" cernment, reflection, and judgment."

PARENT.—Have they acted so basely as this?
Are they so completely abandoned?

DOCTOR.—" It is God's command, that man
" shall not lie with any manner of beast" * * * * *

PARENT.—I must interrupt you here. Surely,
Sir, you are beside yourself! How abominable!
What can you mean by that filthy insinuation? Is
this the language of truth and soberness?

DOCTOR.—" It is God's command not to con-
" taminate the form of the Creator with the brute
" creation. Various beastly diseases common to
" cattle, have appeared amongst the human species
" since the introduction of the Cow-pox. Its ve-
" hement advocates are seriously admonished to
" repent in time, and appeal to Heaven for mercy.
" They should repeat our confession, in the Prayer-
" Book, ' *We have erred and strayed from thy ways*
" ' *like lost sheep : we have followed too much the*
" ' *devices and desires of our own hearts : we have*
" ' *offended against thy holy laws : we have left un-*
" ' *done those things which we ought to have done ;*
" ' *and we have done those things which we ought*
" ' *not to have done ; and there is no health in us.*'
" The vaccinators have been flying in the face of
" HEAVEN, by introducing a beastly disease ; and
" HEAVEN ' *holds them in derision,*—the LORD *laugh-*
" ' *eth them to scorn.*' The Cow-pox, produced by

" presumptuous, impious man, is a daring and pro-
" fane violation of our HOLY RELIGION."

PARENT.—I am really shocked and grieved, and
ashamed for you, Dr. Bragwell, to see how you jest
with serious things ; and how unblushingly you pro-
stitute the words of Sacred Writ ! I cannot possibly
believe you have the smallest reverence for the
DIVINE AUTHOR of that HOLY RELIGION
by which, perhaps, you may profess to be guided.
Be assured, Sir, you will speedily have cause to repent of your unbecoming levity.

DOCTOR.—" Whether vaccination be agreeable
" to the will and ordinances of GOD, is a question
" worthy the consideration of the contemplative and
" learned Ministers of the Gospel of JESUS CHRIST."

PARENT.—Again, let me entreat you to desist
from trifling with that sacred name for such ignoble
purposes. Beware, lest the guilt which you would
thus impute to others, should soon, very soon, hea-
vily recoil on yourself. As one of the " Ministers of
" the Gospel," I now warn and admonish you with
sincerity.

DOCTOR.—" It is the age of insane projects and
" poisonous experiments ; and augmenting evils
" ought to be spiritedly opposed. It is enough to
" excite the most lively indignation, to see human
" beings become the devoted victims to wanton pro-
" jects ; and sometimes to the idle, visionary con-
" ceits of youthful inexperience, pride, and obsti-
" nacy. I never intended to interfere, had not the

" present causes rendered it highly necessary for the
" *salvation* of society."

PARENT.—I fervently pray to ALMIGHTY GOD,
whose name you so often take in vain, that your
own " salvation" may not be hazarded by the preva-
lence of ungovernable passions. Remember the
" mildness and gentleness of Truth."

DOCTOR.—" If the most *daring violation of*
" *truth and reason* had not presented itself in the
" form of vaccination : if the flock had pursued their
" vaccination project with DIFFIDENCE, with MO-
" DESTY, with that DECORUM which distinguishes
" sagacity from obdurate folly, they might have
" formed a superficial vaccinating sect, like the trans-
" fusers of beastly blood into human constitutions ;
" they might have preached *salvation to the faithful*
" in their elaborate sermons, however they had de-
" luded the unwary."

PARENT.—Oh! fie, fie, Dr. Bragwell. Talk
no longer about " *salvation*," and preaching to the
" *faithful*." I once more admonish you ; and trust
you will not again offend against Christian morals
and religion. Where is that " *diffidence*," that " *mo-*
" *desty*," that " *decorum*," of which you speak ?

DOCTOR.—" Irrational projects ! Visionary con-
" ceits ! Obstinate perseverance in error ! Uncon-
" trolled arrogance ! Insane pretensions and delu-
" sions ! The most daring, futile, and ridiculous
" violations of truth and reason ! Cow-pox devas-
" tation ! Exterminating project ! Murderous evil !
" Cruel vaccination ! Bloody contest ! Man is little

" short of insanity, while the raging fury continue!
" When infatuation, intemperate zeal, or wild en-
" thusiasm seize the human mind; CLEAR CONCEP-
" TIONS, TRUE REASONING, AND SOLID JUDGMENT,
" vanish."

PARENT.—So I plainly perceive. I heartily wish,
Dr. Bragwell, you would bring your evidence against
the Cow-pox to a speedy conclusion. What are the
ravings of madmen to me? I want rather to hear the
words of soberness and truth.

DOCTOR.—" My own history of medicine shews
" much of sects, sectarists, false systems, &c. with
" the *raving-mad* conceits of every age; and advises
" the admission of NOTHING BUT POSITIVE TRUTH.
" We have seen the world, and some of the faculty,
" quicksilver-mad—Viper-broth mad—Tincture of
" Cantharides mad — Tar-water mad — Stevens's
" stone-powder mad—Calve's pluck-water mad—
" Digitalis mad—Arsenic mad—Sugar of lead mad
" —Magnetism mad—Caustic bougies mad—Me-
" phitic-water mad—Electricity and galvanism mad—
" Hemlock mad—Salt-water mad—Portland powder
" mad—Acid and alcali mad—Alchemistic mad—
" Transmutatio metallorum into gold mad—Gas and
" vital air mad—Poison mad—Nitric acid mad—
" Phosphorus mad—Vegetable diet mad—Le fevre
" gout mad—Henbane or night-shade mad—Me.
" tallic tractor mad—Iced-water mad—Jalap and
" calomel mad—Buzaglo mad—Bleeding and butter-
" milk mad—COW-POX mad.

" All these strange conceits and impositions have
" been supported by *knavery, ignorance, folly,* and
" *false faith.*"

PARENT.—Pray, Dr. Bragwell, did you ever treat
on *small-pox madness?* I wish it may not be your
own disorder. At least I must confess, that the
flights and aberrations of your enlightened mind,
are far too lofty and eccentric for my plain under-
standing! Your statements produce no conviction,
either of the merits or demerits of Cow-pox inoculation. I remain just as ignorant on the subject as
before I consulted you; and should much wish to
know, whether any examination of adverse facts,
or supposed failures, has been set on foot by certain
vaccinating societies in London?

DOCTOR.—" It is a degrading circumstance to
" vaccination, that the PRINCIPAL PROMOTERS of the
" *vile fallacy* hide themselves behind the curtain;
" and force out raw, undisciplined, simple, unlet-
" tered recruits,—or eccentric, wild-headed, versa-
" tile, superficial characters,—to defend their de-
" tected Cow-pox deceptions, to put a good face on
" a desperate cause. Vaccination cunning would be
" the most superlative, refined, and imposing; if
" vaccinators had sufficient mental powers to hide
" their cunning. Have these vaccinators instituted
" inquiries? No.—Have they sought for opposing
" truths? No.—Have they not endeavoured vio-
" lently to suppress the truth? Yes.—It is known,
" that they have *never* honourably promoted any re-

" gular system of inquiry.   Why do not the FORMER
" SUPPORTERS OF DIVINE JENNERIAN VACCINATION
" come out collectively, and declare they are as
" much as ever convinced of its utility? The Royal
" College of Surgery, the true judges on the sub-
" ject, were never consulted ; and, to the immortal
" honour of our Royal College of Physicians, they
" did not countenance vaccination in an unqualified
" manner."

PARENT.—Indeed, Doctor Bragwell! Do none
of the physicians and surgeons who encourage vaccination belong to the two Royal Colleges of London? Or, do you only intimate that they were
not consulted in a body? Perhaps, as Collegiates, they
knew nothing of the matter ; or (at the period
you allude to) had seen but little of the progress and
effects of vaccination, and therefore could not be
expected, *as a body*, to give any collective opinion.
I suppose you do not deny that a large number of
individuals, who are members of the two Colleges,
have in their private capacity given the most decided
encouragement to vaccination. Can you deny this,
Doctor? I ask for information, being but imperfectly
acquainted with the names of those experienced and
learned gentlemen of the profession.

Upon recollection, however, I think you must be
misinformed, in some degree, about the opinion of
the College of Physicians ; for I have seen the copy
of a letter, signed by Dr. Gisborne, as President of
that College, and sent to the Committee of the

3

House of Commons (April 13, 1802), in which that
learned body gave their decided " COUNTENANCE TO
" VACCINATION," in the following words:—viz.
that the Vaccine Inoculation, when properly con-
ducted, was perfectly safe, and highly deserving the
encouragement of the public. And, as to the " prin-
" cipal promoters," or the " former supporters" of
vaccination " hiding themselves behind the curtain,"
and " *never*" instituting " a regular system of in-
" quiry," as you say ; that is a point on which I
should not have felt competent to return a word of
answer, if the visitor, who has this moment entered
the room, had not shewn me a printed account of
the result of some investigations, dated the 2d of
January 1806. But give me leave to ask, Who do
you mean by the " FORMER SUPPORTERS and PRIN-
" CIPAL PROMOTERS *of the vile fallacy ?*"

DOCTOR.—A very fair question, Reverend Sir ;
a very fair question, and soon answered. You must
know, then, that " the ARTFUL erect the structure of
" their extraordinary success on a supposition, that
" the majority of mankind are absolute fools, cre-
" dulous idiots, and easily seduced."

PARENT.—True, Doctor, true : but I hope the
individuals of whom you now speak, the respectable
" former supporters and principal promoters of di-
" vine Jennerian vaccination," are not so completely
abandoned and *artful* as this ? Pray what are their
names? Where are they to be found ?

E

DOCTOR.—" Cow-pox deception! A vile fal-
" lacy! A desperate bad cause! My certain, expe-
" rienced, and successful mode of inoculating for
" the small-pox, becomes necessary from cow-pox
" failure.   Out of many thousands, nay MILLIONS,
" scarcely ANY ONE has died from small-pox ino-
" culation!"

PARENT.—It is very unfortunate that my own
dear babe, my first-born child, should be an exception to this universal success of the small-pox
inoculation: but, Sir, I am asking you to tell me
the names of the first promoters, the principal supporters, of the vaccine practice; who now retreat
from the field of battle, and " force out raw, un-
" disciplined, simple, unlettered recruits?"

DOCTOR.—See, then, in my last book; turn to
page 93, and you will find who they were.   My own
list was " extracted from the learned Dr. Moseley's
" pamphlet," and his was taken from " an adver-
" tisement in a daily paper of July 19th, 1800;"
but, a similar one appeared in all the medical journals
of that period.   The original was as follows:

" Many unfounded reports having been circulated,
" which have a tendency to prejudice the mind of
" the public against the inoculation of the Cow-pox;
" we the undersigned physicians and surgeons, think
" it our duty to declare our opinion, that those per-
" sons who have had the Cow-pox are perfectly se-
" cure from the future infection of the small-pox.

" We also declare, that the inoculated Cow-pox

" is a much milder and safer disease than the inocu-
" lated small-pox.

" William Saunders, M. D.	Edward Ford.
Matthew Baillie, M. D.	Astley Cooper.
Henry Vaughan, M. D.	John Abernethy.
Maxwell Garthshore, M. D.	Joseph Hurlock.
John Coakley Lettsom, M.D.	William Blair.
James Sims, M. D.	Samuel Chilver.
John Sims, M. D.	J. M. Good.
William Lister, M. D.	James Horsford.
Robert Willan, M. D.	Francis Knight.
C. Stanger, M. D.	James Leighton.
Alexander Crichton, M. D.	James Moore.
Thomas Bradley, M. D.	Thomas Paytherus.
Thomas Denman, M. D.	Thomas Pole.
John Squire, M. D.	J. W. Phipps.
Richard Croft, M. D.	John Ring.
Robert Batty, M. D.	James Simpson.
R. J. Thornton, M. D.	H. L. Thomas.
Richard Dennison, M. D.	Jonathan Wathen.
Henry Cline.	Thomas Whateley. "

N. B. The signatures which were added to this list after the
19th of July 1800, made the total number *seventy-two.*
See RING's Treatise, Vol. I. p. 298.

PARENT.—Do you think all those physicians and
surgeons were truly honourable men ?

DOCTOR.—" Very respectable practitioners."

PARENT.—And were there any more respectable promoters ?

DOCTOR.—" The respectable practitioners, exa-
" mined by the Honourable Committee of the House

" of Commons, who applauded vaccination"—and whose names I have copied from the ingenious Dr. Moseley's work, were these:

" Dr. Ash.	Mr. Keate.
Mr. Home.	Mr. Robt. Keate.
Dr. Woodville.	Mr. T. Nash.
Dr. Blane.	Mr. Gardner.
Mr. Knight.	Dr. Lister
Rev. G. C. Jenner.	Mr. Cline.
Mr. John Griffiths.	Dr. Bradley.
Mr. Wm. Cuff.	Sir Walter Farquhar.
Dr. Thos. Denman.	Dr. James Sims.
Dr. Croft.	Dr. Saunders.
Sir G. Baker.	Dr. Lettsom.
Dr. G. Pearson.	Dr. Frampton.
Dr. Thornton.	Dr. Baillie."

PARENT.—"A very respectable" list indeed, Doctor; and many of them, I see, are the same who at the first had given their sanction to vaccination, in July 1800. But, do you know of any more " respectable" names of the EARLIEST SUPPORTERS?

DOCTOR.—Why, yes: I have given another long catalogue at page 95 of my book on " Cow-pox " Inoculation no Security, &c." and they are the physicians and surgeons who first of all composed the " Medical Council of the ROYAL JENNERIAN " SOCIETY FOR THE EXTERMINATION OF THE SMALL- " POX, instituted in Salisbury Court, Fleet Street, " January 1803."

PARENT —Favour me with their names, that I may see if there be any among them who originally stood forward in the cause of vaccination. You say these gentlemen formed the " Medical Council;" and therefore were, doubtless, only a small part of the physicians and surgeons supporting this " ROYAL " JENNERIAN SOCIETY." I imagine the members of this Council have been changed occasionally, as in other well-regulated committees ?

DOCTOR.—Yes : I only speak now of the Medical Council for that one year 1803.

" The patrons and philanthropic promoters and
" protectors may be found in the *red book* *, amongst
" whom stands foremost, His most GRACIOUS MA-
" JESTY, the ROYAL FAMILY, the Archbishop of
" Canterbury, the Bishops, Nobility, Gentry, &c.
" &c."

* The "*red-book*" of that date gave an incorrect account of the patrons and directors, &c. of the Royal Jennerian Society : their names, taken from the Society's official register, were then (1803) as follow :

PATRON—THE KING.

PATRONESS—THE QUEEN.

VICE-PATRONS.

His Royal Highness the Prince of Wales.
His Royal Highness the Duke of York.
His Royal Highness the Duke of Clarence.
His Royal Highness the Duke of Cumberland.
His Royal Highness the Duke of Cambridge.
His Royal Highness the Duke of Gloucester.

4

VICE-PATRONESSES.

Her Royal Highness the Princess of Wales.
Her Royal Highness the Duchess of York.
Her Royal Highness the Princess Sophia Augusta.
Her Royal Highness the Princess Elizabeth.
Her Royal Highness the Princess Mary.
Her Royal Highness the Princess Sophia.
Her Royal Highness the Princess Amelia.

PRESIDENT.

His Grace the Duke of Bedford.

SUB-PATRONESSES.

Duchess of Devonshire.
Duchess of Marlborough.
Duchess of Rutland.
Dss. of Northumberland.
Marchioness of Bath.
Marchioness of Hertford.
Marchioness of Bute.
Marchioness of Sligo.
Countess of Carlisle.
Countess of Sutherland.
Countess of Dartmouth.
Countess Fitzwilliam.
Countess Spencer.
Countess Bathurst.
Countess of Uxbridge.
Countess Grosvenor.
Countess Camden.
Countess of Carnarvon.
Countess of Darnley.
Viscountess Lowther.
Rt. Hon. Lady Garlies.
Rt. Hon. Lady Harvey.
Viscountess St. Asaph.

Lady Theodosia Maria Viner.
Lady Willoughby de Eresby.
Lady Hobart.
Lady Sherborne.
Lady Rous.
Lady Louisa Brome.
Lady Mary Stopford.
Lady Sheffield.
Lady Templeton.
Lady Huntingfield.
Lady Frances Moreton.
Lady Caroline Wrottesley.
Hon. Mrs. Harcourt.
Hon. Mrs. Spencer Perceval.
Lady Richard Carr Glyn.
Lady Nepean.
Mrs. Beaumont.
Mrs. Boucherett.
Mrs. Burdon.
Mrs. Chaplin of Blanckney.
Mrs. Chute.
Mrs. Charles Grant.
Mrs. Henry Hicks.

Mrs. Thomas Kinscote.
Mrs. Beeston Long.
Mrs. Manning.
Mrs. Neave.

Mrs. Thellusson.
Mrs. Charles Wall.
Miss Angerstein.

VICE-PRESIDENTS.

Archbishop of Canterbury.
Lord Chancellor.
Duke of Somerset.
Duke of Devonshire.
Duke of Northumberland.
Marquis of Hertford.
Marquis of Worcester.
Earl of Shrewsbury.
Earl of Derby.
Earl of Westmoreland.
Earl of Berkeley.
Earl of Egremont.
Earl of Harcourt.
Earl of Hardwicke.
Earl Spencer.
Earl of Liverpool.
Earl St. Vincent.
Earl of Darnley.
Earl of Limerick.
Earl Moira.
Lord Viscount Melville.
Bishop of London.
Bishop of Durham.
Lord Viscount Castlereagh.
Lord Pelham.
Lord Somerville.
Lord Rous.
Lord Carrington.
Lord Gwydir.
Lord Auckland.

Lord Hobart.
Lord Gardner.
Right Hon. Charles Abbot, Speaker of the House of Commons.
Right Hon. Charles Price, Lord Mayor, M. P.
Right Hon. Henry Addington, M. P.
Rt. Hon. Wm. Pitt, M. P.
Rt. Hon. Sir Joseph Banks, Bart. K. B.
Hon. Admiral Berkeley, M. P.
Hon. Ch. J. Fox, M. P.
Hon. Charles Grey, M. P.
Sir Henry Mildmay, Bart. M. P.
Sir Francis Baring, Bart. M. P.
Sir John Wm. Anderson, Bart. M. P.
Sir Wm. Curtis, Bart. M. P.
Edward Jenner, M. D.
John Julius Angerstein, Esq.
Thomas Bernard, Esq.
Thomson Bonar, Esq.
Harvey Ch. Combe, Esq. M. P.
John Fuller, Esq. M. P.
Abraham Goldsmid, Esq.
George Hibbert, Esq.
W. S. Poyntz, Esq. M. P.
R. B. Sheridan, Esq. M. P.

PARENT.—Who were the gentlemen at that time composing the Medical Council?

DOCTOR.—They were as follow :

" PRESIDENT, EDWARD JENNER, M. D.

" VICE-PRESIDENT, *my old and worthy Friend, the* " *Friend of all human Society*, J. C. LETTSOM, M. D.

" William Babington, M. D.
" Robert Batty, M. D.
" Gilbert Blane, M. D.
" Thomas Bradley, M. D.
" Isaac Buxton, M. D.
" John Clark, M. D.
" Alex. Crichton, M. D.
" Richard Croft, M. D.
" Thomas Denman, M. D.
" W. Pitts Dimsdale, M. D.
" Philip Elliot, M. D.
" Sir Walter Farquhar, Bart. M. D.
" W. M. Fraser, M. D.
" James Hamilton, M. D.
" William Hamilton, M. D.
" William Hawes, M. D.
" Robert Hooper, M. D.
" A. J. G. Marcet, M. D.
" Samuel Pett, M. D.
" Richard Powell, M. D.
" James Sims, M. D.
" William Lister, M. D.
" Joseph Skey, M. D.
" Thomas Turner, M. D.
" Robert Willan, M. D.

" John Abernethy, Esq.
" John Addington, Esq.
" C. R. Aikin, Esq.
" William Chamberlaine, Esq.
" Henry Cline, Esq.
" Astley Cooper, Esq.
" John Curtis, Esq.
" John Dimsdale, Esq.
" Edward Ford, Esq.
" Joseph Fox, Esq.
" William Gaitskell, Esq.
" John Griffiths, Esq.
" Everard Home, Esq.
" Joseph Hurlock, Esq.
" Charles Johnson, Esq.
" George Johnson, Esq.
" Thomas Key, Esq.
" L. Leese, Esq.
" J. Pearson, Esq.
" J. Ring, Esq.
" James Upton, Esq.
" Allen Williams, Esq. Secretary to the Medical Council.
" John Walker, M. D."

PARENT.—Again I observe many of the original supporters, Dr. Bragwell: the same individuals who at first, in 1800, resisted the " unfounded reports " in circulation," still continue their sanction, and probably gave their money as subscribers to the society. What, give their time, their money, their names, and reputation ; all for nothing, Dr. Bragwell ? They must indeed be greatly in earnest. I perceive then, that these disinterested professional gentlemen well merit the title you give them of " PHILANTHROPIC PROMOTERS AND PROTECTORS ;" and that they are deservedly sanctioned by " HIS " MOST GRACIOUS MAJESTY, THE ROYAL FAMILY, " THE ARCHBISHOP OF CANTERBURY, THE BISHOPS, " NOBILITY, GENTRY, &c. &c."

I begin now to suspect, you may have unjustly stigmatized, the cause of vaccination, Dr. Bragwell. How is it that we find such royal, noble, reverend, and dignified characters, in company with so many experienced physicians and surgeons, for the encouragement of what you have emphatically denominated " A VILE FALLACY ?" Do you not hazard incurring the imputation of *vilifying* your superiors ? And are you quite sure that ALL these " *very re-* " *spectable*" medical gentlemen have retired " be- " hind the curtain," as if ashamed of the Jennerian practice ? This is a very grave, momentous, and necessary question, Dr. Bragwell ! Either your character or theirs must infalliby suffer ; and it will soon be discovered by the scrutinizing eye of the world,

who is " ACTING IN DIRECT VIOLATION OF
" TRUTH." Recollect, Sir, what you and I have
before remarked, concerning the sacred and invio-
lable obligations of TRUTH ; for, if it should hereafter appear that YOU have " *perverted the truth*,"
whatever be your motive, rest assured that you can
never, never wipe off the odium you will justly incur.
May Heaven avert the punishment such base and
diabolical conduct would deserve ! ! !

DOCTOR.—" The opposers of Cow-pox are
" ready to meet its advocates, being secure of a glo-
" rious victory ! It is known that the vaccinators
" (as I said before) have never promoted any system
" of inquiry."

PARENT.—I will take the liberty, in reply to
your bold and unqualified assertion on this point, to
read the printed paper which I hold in my hand. It
appears to have been published by THE ROYAL
JENNERIAN SOCIETY FOR THE EXTER-
MINATION OF THE SMALL-POX.

AT a special meeting of the Board of Directors, held
at the central house of the Society, No. 14, Salisbury Square, Fleet Street, the Report of the Medical Council, on the subject of Vaccine Inoculation,
having been laid before the Board,

Resolved,—That the same be immediately printed
under the direction of the Medical Council, and that
they be requested to subjoin their individual signatures to the Report for publication.

Extract from the Minutes. CHARLES MURRAY, Sec.

REPORT.

The Medical Council of the Royal Jennerian Society, having been informed that various cases had
occurred, which excited prejudices against Vaccine
Inoculation, and tended to check the progress of
that important discovery in this kingdom, appointed
a committee of twenty-five of their members to inquire, not only into the nature and truth of such
cases, but also into the evidence respecting instances
of small-pox, alleged to have occurred twice in the
same person.

In consequence of this reference, the Committee
made diligent inquiry into the history of a number of
cases, in which it was supposed that vaccination had
failed to prevent the small-pox, and also of such cases
of small-pox, as were stated to have happened sub-
sequently to the natural or inoculated small-pox.

In the course of their examination the Committee
learned, that opinions and assertions had been advanced and circulated, which charged the Cow-pox
with rendering patients liable to particular diseases,
frightful in their appearance, and hitherto unknown;
and judging such opinions to be connected with the
question as to the efficacy of the practice, they
thought it incumbent upon them to examine also
into the validity of these injurious statements respecting vaccination.

After a very minute investigation of these subjects,

the result of their inquiries has been submitted to the Medical Council ; and from the Report of the Committee it appears :

I. That most of the cases, which have been brought forward as instances of the failure of vaccination to prevent the small-pox, and which have been the subjects of public attention and conversation, are either wholly unfounded or grossly misrepresented.

II. That some of the cases are now allowed, by the very persons who first related them, to have been erroneously stated.

III. That the statements of such of those cases as are published, have, for the most part, been carefully investigated, ably discussed, and fully refuted, by different writers on the subject.

IV. That notwithstanding the most incontestable proofs of such misrepresentations, a few medical men have persisted in repeatedly bringing the same unfounded and refuted reports, and misrepresentations, before the public ; thus perversely and disingenuously labouring to excite prejudices against vaccination.

V. That in some printed accounts adverse to vaccination, in which the writers had no authenticated facts to support the opinions they advanced, nor any reasonable arguments to maintain them, the subject has been treated with indecent and disgusting levity ; as if the good or evil of society were fit objects for sarcasm and ridicule.

VI. That when the practice of vaccination was first introduced and recommended by Dr. Jenner, many persons, who had never seen the effects of the vaccine fluid on the human system, who were almost wholly unacquainted with the history of vaccination, the characteristic marks of the genuine vesicle, and the cautions necessary to be observed in the management of it, and were therefore incompetent to decide whether patients were properly vaccinated or not, nevertheless ventured to inoculate for the cow-pox.

VII. That many persons have been declared duly vaccinated, when the operation was performed in a very negligent and unskilful manner, and when the inoculator did not afterwards see the patients, and therefore could not ascertain whether infection had taken place or not ; and that to this cause are certainly to be attributed many of the cases adduced in proof of the inefficacy of cow-pox.

VIII. That some cases have been brought before the Committee, on which they could form no decisive opinion, from the want of necessary information as to the regularity of the preceding vaccination, or the reality of the subsequent appearance of the small-pox.

IX. That it is admitted by the Committee, that a few cases have been brought before them, of persons having the small-pox, who had apparently passed through the cow-pox in a regular way.

X. That cases, supported by evidence equally strong, have been also brought before them, of persons who, after having once regularly passed through the small-pox, either by inoculation or natural infection, have had that disease a second time.

XI. That in many cases, in which the small-pox has occurred a second time, after inoculation or the natural disease, such recurrence has been particularly severe, and often fatal; whereas, when it has appeared to occur after vaccination, the disease has generally been so mild, as to lose some of its characteristic marks, and even sometimes to render its existence doubtful.

XII. That it is a fact well ascertained, that, in some particular states of certain constitutions, whether vaccine or variolous matter be employed, a local disease only will be excited by inoculation, the constitution remaining unaffected; yet that matter taken from such local vaccine or variolous pustule is capable of producing a general and perfect disease.

XIII. That if a person, bearing the strongest and most indubitable marks of having had the small-pox, be repeatedly inoculated for that disease, a pustule may be produced, the matter of which will communicate the disease to those who have not been previously infected.

XIV. That, although it is difficult to determine precisely the number of exceptions to the practice, the Medical Council are fully convinced that the failure of vaccination, as a preventive of the small-pox, is a *very rare* occurrence.

XV. That of the immense number who have been
vaccinated in the army and navy, in different parts
of the United Kingdom, and in every quarter of the
globe, scarcely any instances of such failure have
been reported to the Committee, but those which
are said to have occurred in the metropolis, or its
vicinity.

XVI. That the Medical Council are fully assured,
that in very many places, in which the small-pox
raged with great violence, the disease has been speedily and effectually arrested in its progress, and in
some populous cities wholly exterminated, by the
practice of vaccination.

XVII. That the practice of inoculation for the
small-pox, on its first introduction into this country,
was opposed and very much retarded, in consequence
of misrepresentations and arguments drawn from as-
sumed facts, and of miscarriages arising from the
want of correct information, similar to those now
brought forward against vaccination ; so that nearly
fifty years elapsed before small-pox inoculation was
fully established.

XVIII. That, by a reference to the bills of mortality, it will appear that, to the unfortunate neglect
of vaccination, and to the prejudices raised against it,
we may, in a great measure, attribute the loss of
nearly two thousand lives by the small-pox, in this
metropolis alone, within the present year.

XIX. That the few instances of failure, either in
the inoculation of the cow-pox, or of the small-pox,

ought not to be considered as objections to either practice, but merely as deviations from the ordinary course of nature.

XX. That if a comparison be made between the preservative effects of vaccination, and those of inoculation for the small-pox, it would be necessary to take into account the greater number of persons who have been vaccinated within a given time : as it is probable, that, within the last seven years, nearly as many persons have been inoculated for the cow-pox, as were ever inoculated for the small-pox, since the practice was introduced into this kingdom.

XXI. That, from all the facts which they have been able to collect, it appears to the Medical Council, that the cow-pox is generally mild and harmless in its effects ; and that the few cases, which have been alleged against this opinion, may be fairly attributed to peculiarities of constitution.

XXII. That many well-known cutaneous diseases, and some scrofulous complaints, have been represented as the effects of vaccine inoculation, when in fact they originated from other causes, and in many instances occurred long after vaccination ; and that such diseases are infinitely less frequent after vaccination, than after either the natural or inoculated small-pox.

Having stated these facts, and made these observations, the Medical Council cannot conclude their report upon a subject so highly important and

G

interesting to all classes of the community, without making this *solemn Declaration*;

THAT, IN THEIR OPINION, FOUNDED ON THEIR OWN INDIVIDUAL EXPERIENCE, AND THE INFORMATION WHICH THEY HAVE BEEN ABLE TO COLLECT FROM OTHERS, MANKIND HAVE ALREADY DERIVED GREAT AND INCALCULABLE BENEFIT FROM THE DISCOVERY OF VACCINATION: AND IT IS THEIR FULL BELIEF, THAT THE SANGUINE EXPECTATIONS OF ADVANTAGE AND SECURITY, WHICH HAVE BEEN FORMED FROM THE INOCULATION OF THE COW-POX, WILL BE ULTIMATELY AND COMPLETELY FULFILLED.

(Signed)

Edward Jenner, M. D. President of the Council.
J. C. Lettsom, M. D. V. P.
John Ring, V. P.
Joseph Adams, M. D.
John Addington.
C. R. Aikin.
Wm. Babington, M. D.
M. Baillie, M. D.
W. Blair.
Gil. Blane, M. D.
Isaac Buxton, M. D.
Wm. Chamberlaine.

John Clarke, M. D.
Astley Cooper.
Wm. Daniel Cordell.
Richard Croft, M. D.
Tho. Denman, M. D.
John Dimsdale.
Henry Field.
Edward Ford.
Joseph Fox.
Wm. M. Fraser, M. D.
William Gaitskell.
William Hamilton, M. D.
John Hingeston.

Everard Home.

Robert Hooper, M. D.

Joseph Hurlock.

John Jones.

Thomas Key.

Francis Knight.

E. Leese.

L. Leese.

William Lewis.

William Lister, M. D.

Alex. Marcet, M. D.

Joseph Hart Myers, M. D.

Jas. Parkinson.

Thos. Paytherus.

John Pearson.

George Rees, M. D.

John Gibbs Ridout.

J. Squire, M. D.

Jas. Upton.

J. Christian Wachsell.

Thomas Walshman, M. D.

Robert Willan, M. D.

Allen Williams.

James Wilson.

J. Yelloly, M. D.

January 2, 1806.

John Walker, *Secretary to the Council.*

PARENT.—Tell me now, *solemnly* and *conscientiously*, Dr. Bragwell, what you think of this REPORT, and of the signatures affixed to it ? I still observe many of the former promoters, the principal supporters of vaccination, whom you already have confessed to be " *very respectable practitioners.*" Do any of these fifty-one physicians and surgeons belong to either of the Royal Colleges in London ? Perhaps they are only a troop of your " RAW RECRUITS." Be so candid as to answer me, Dr. Bragwell.

DOCTOR.—Indeed, Reverend Sir, you are very pointed and arrogant in asking me so many questions. Have I not peremptorily and decidedly told you before, that " *if vaccination spoke the* TRUTH *it would* " *not be believed.* For, who will credit any enthu- " siastic vaccinator ? Who will believe a word they

" say ? None ! unless they hate the truth and
" cherish deception ! Can any one expect truth from
" the fabricators of dreams, of visions ? from the
" promulgators of prejudices ? None, but the cre-
" dulous, and cow-pox maniacs ! None, but those
" who are no judges of physic !"

PARENT.—Then, most learned Sir, are *you* the
only judge of physic ; the only professional man devoid of prejudices ; the only one who never has been
overcome by dreams ; the only person who loves
truth, and hates deception ? You do not, it appears,
much relish my pointed questions ; and you take
good care to answer very few of them. I am, therefore, compelled to suppose what replies you would
give, if the truth, the whole truth, and nothing but
the truth, were told. Under these circumstances,
Doctor Bragwell, it must be loss of time, both to
you and me, to make any farther inquiries. I shall
now retire, and meditate upon your oracular suggestions ! The gentleman who sits near me, will bear
witness that I have given you a fair opportunity of
justifying yourself and pleading your own cause. My
opinion is at length made up, on the subject of vaccination ; and unless I shall meet with opponents to
the new practice, of a more argumentative and honest turn of mind than you, Doctor Bragwell, I must
yield to my present determination of inoculating my
own children, and of giving ALL POSSIBLE ENCOU-
RAGEMENT TO THE COW-POX THROUGHOUT MY
WHOLE PARISH.

SURGEON.—Reverend Sir, I greatly approve of your determination, and I bear witness this day to the ingenuousness and candour of your deportment towards Doctor Bragwell; of whom, and to whom, I am at present come to say a few words. I am a Surgeon, residing in London, and well known to——

DOCTOR.—Speak on, Sir; speak on; I am the sworn advocate of truth; and if I cannot convince the Reverend Gentleman, it is not for want of pains-taking and zeal on my part. I am not ashamed of what I have said; and have told him nothing that has not been already published to all the world, in the third edition of my unanswerable work, or in my other practical writings. Not a sentence, Sir, nor a single word, do I retract. Come Sir, come Sir, speak on. You, Mr. Surgeon, have you too joined these " *violent enthusiasts?*" You, of whom I had hoped better things ? Why, Sir, you have been a strenuous defender of the truth, in former times: it was you, chiefly, who confronted and silenced the PHILO-ACIDI, a few years ago; and I will presently either silence you, or chase you from my doors, as I advise all persons to do with incorrigible vaccinators. " I " am happy to say, I never in one instance recom- " mended vaccination;" and I have told this Cler-gyman, that " the opposers of cow-pox are ready " to meet its advocates, being secure of a glorious " victory."

SURGEON.—Be assured that if ever you *meet* them, it will be to your confusion and disgrace : but,

such a meeting is needless; for, your heap of anti-
vaccinarianism is before the public. You have raked
together into one mass (a disgusting collection !) all
the conjectures or hearsay tales, all the pretended
cases of failure or disaster, which any person would
communicate for the gratification of your vanity ;
with all the adverse facts hitherto published by others,
though answered again and again. So that, whoever
gives you, Doctor Bragwell, a proper reply, will have
refuted the objections of every adversary that has appeared in this combat.

DOCTOR.—Well, well, Sir ; go on with your
invective. " I have no time for controversy." The
TRUTH lies in a narrow compass. You have, of
course, read my collection of FACTS, and examined
them attentively. Very true; Mr. Surgeon ; in refuting me, you will vanquish the whole host of *Anti-
vaccinarians*. Did you ever hear of the " ANTI-
VACCINARIAN SOCIETY ?"

SURGEON.—I am happy to have an opportunity
of telling you, and this Reverend Gentleman, (whose
patience equals his liberality of sentiment,) that your
" MATTERS OF FACT," and the " ANTI-VAC-
" CINARIAN SOCIETY," of which you are known
to be the *fac totum*, disgrace the medical profession ;
and impeach the moral characters of men of all ranks
in the Faculty, who are far, very far your superiors
in reputation and virtue. Your publication is a
gross libel (I repeat it) upon professional gentlemen
of the first name and celebrity, and merits no other

epithet than a most scandalous vehicle of falsehoods, tending to the worst consequences in society. You pique yourself on being the leader and father of the " Anti-vaccinarian Society :" and it is no wonder you should foster this darling of your own begetting. *Doctor Squirrell* (alias *Mr. S. the Apothecary*) is worthy of being your colleague in this establishment ; which, indeed, looks too much like a mercenary contrivance, to bring the dissatisfied patients to your shop ! It is perfectly of a piece with your self-commendatory puffs in the daily newspapers,—your multiform circular hand-bills,—your numerous placards on the dead-walls of the town,—your acknowledged " public exhibitions in the lecture-room,"—your " cart-loads of children in Saville Row,"—your insidious bribes, feastings, and humiliating attentions to Apothecaries, &c. who have been artfully seduced to countenance your plans !

PARENT.—Oh ! does the learned Oxonian stoop so low as all this ? Can he, then, be respected among his medical associates ? But, I don't understand what kind of an institution Doctor Bragwell's " *Anti-vac-* " *cinarian Society*" is, of which he makes his boast.

SURGEON. *(aside.)*—Reverend Sir, you seem not to know any thing of the estimation in which this Physician is held, among really honourable men in London. I say nothing about his underlings, or the self-dubbed Doctors who have enlisted into his service. You may conjecture what kind of an establish-

ment his " Anti-vaccinarian Society" is, by an ad-
vertisement I shall now read for your information :

" 1806.—THE ANTI-VACCINARIAN SOCIETY gra-
" tuitously purpose to examine all accounts and facts
" of cow-pox failure ; whether of small-pox, or
" beastly breakings out after cow-pox inoculation :
" they therefore respectfully entreat the faculty of
" physic, and the public in general, to send whatever
" they know of ill consequences arising from cow-
" pox to Dr. ROWLEY, No. 21, Saville Row.   The
" Society likewise humanely intend to inoculate the
" poor with small-pox MILD MATTER, according to
" Dr. ROWLEY's directions, *gratis*; and they hope
" for assistance from the Nobility, Gentry, and
" Public in general, to second their efforts *in sup-*
" *porting the* TRUE *cause of humanity against cow-*
" *pox injuries.   The tyranny, the cruel despotic ty-*
" *ranny, of forcing cow-pox misery on the innocent*
" *babes of the poor, whether they will or not, is con-*
" *sidered* A GROSS VIOLATION OF RELIGION, MO-
" RALITY, LAW, AND HUMANITY ; *and the sooner*
" *suppressed, the better for mankind.*"—But the
cream of this joke is, that the worthy Oxonian and
his adherents have found it no easy matter to get a
Bookseller, or Publisher, sufficiently prejudiced, or
short-sighted, to concur in their designs ; for, says
the Advertiser, in conclusion, " *Some of the public*
" *prints, and even Booksellers, have denied* (i. e. re-
" fused) *publishing or vending this mass of evidence*
" *against cow-pox,*"—doubtful whether the FACTS

and TRUTHS therein contained are capable of proof, or were not such as would involve all the parties concerned in utter ruin ! This hint tends to confirm my settled opinion, both of the " Anti-Vaccinarian'' deceptions, and of certain Booksellers' inflexible honesty.

DOCTOR.—While you are nibbling and haggling at extraneous *minutiæ*, I am for deciding the whole case by an appeal to matters of fact : Sir, let me repeat, what I have told this Reverend Visitor, " FACTS are " STUBBORN THINGS, AND MUST ULTIMATELY PRE- " VAIL." Have you, or have you not, deliberately and honestly examined my " *matters of fact,*" my " *visible, indubitable, and self-evident facts,*" the great " *mass of evidence against Cow-pox*" which is collected at the end of my third edition ? Have you, or have you not, Sir ? I demand a plain, categorical reply, in the presence of this Clergyman, who seems rather leaning towards the side of prejudice already. Tell me, Sir ; tell me : have you investigated this whole body of evidence, both internal and external, and weighed its intrinsic worth ? This is the question.

SURGEON.—Do you mean ALL your alleged facts and cases of failure ? I believe they are not ALL authentic ; notwithstanding it was impossible for me, or any one person, to inquire into every circumstance as it regards them ALL. Shall I explain my objections more in detail ?

DOCTOR.—I tell you ALL, ALL are true ; and

H

I have declared to this Clergyman that " I am ready
" to meet publicly the whole host of vaccinators, or
" their protectors, to *demonstrably* PROVE ALL
" THE ALLEGATIONS promulgated in my book
" TRUE." I can not only prove them, Sir, but
prove them *demonstrably*.

SURGEON.—You need not, as I suggested, take
the trouble of convening these numerous partizans ;
you now " *meet*" me, and can demonstrate the truth
of your allegations as easily before two persons, as
before two thousand. I am willing to put the whole
question to this issue : if I do not " *demonstrably*
" *prove*" some of your representations to be " *visibly,*
" *indubitably, and self-evidently*" FALSE, I will yield
the palm of victory to the Anti-Vaccinators, and
" let cow-poxing be banished from the face of the
" earth ;" but if I prove this clearly and palpa-
bly, from internal evidence, from no other evidence
than your own book affords, it will be granted that
I have little reason to take your bare word for the
truth of the remaining allegations. By this com-
promise, you see, I have taken upon myself the
onus probandi, and will leave your book to speak
for itself. This must lessen your trouble very much,
and save a great deal of time : for, it is more tedious
to examine the whole, than only a small part of
your statement ; which includes, as you tell us in
great triumph, " FIVE HUNDRED AND FOUR PROOFS
" OF FAILURE."

DOCTOR.—" A glorious victory," Sir, for the

" Anti-vaccinarians!" Here, Sir, here; take this copy of my work, "THE THIRD EDITION, PRINTED FOR " THE AUTHOR, BY J. BARFIELD, WARDOUR STREET, " LONDON, 1806." Observe, this is the edition I appeal to. There might, perhaps, have been a few trifling inaccuracies in the two former editions.

SURGEON.—FIRST, Dr. Bragwell, be pleased to consult page 129, where I read thus: " The five " hundred and four facts against the continuance of " Cow-pox inoculation, out of which seventy-five " have died devoted victims to the filthy project, " are but as a drop of water added to the sea." Compare these words with the following, at page 98 : " There have appeared vaunting threats of refuting " the two hundred and eighteen cases ; but there are " now between four and five hundred cases in this " *second* edition, which must likewise be refuted ; " but unless the refuting vaccinators can bring the " fifty and more to life who have perished, who have " died under vaccination, or by small-pox, or beastly " diseases after cow-poxing, &c."

As your readers will form different opinions concerning the cause of this inconsistency in your statements, I need not attempt an explanation ; but shall proceed to consider which comes nearest the truth, whether the five hundred and four cases, or the smaller number of failures. I take for granted, however, that your sum total, at the bottom of Case 504, is intended to be received as the true account ; viz. " *small-pox after cow-pox, &c. five hundred and four :*

H 2

" *died, seventy-five*"—although this round assertion does not quite agree with your words at page 101;
namely, " in the foregoing cases, there are *four*
" *hundred and forty* instances of small-pox after
" cow-pox."   But, on counting over your cases, and
examining their particulars, I find more than one
hundred and fifty events which you say arose from
other causes than the small-pox ; and therefore de-
ducting upwards of one hundred and fifty from five
hundred and four, the result will be nearly according to the last mentioned account.

Out of these five hundred and four patients, there
are not less than *one hundred and twenty-seven*
whose names or places of abode are wanting ; and
two hundred and thirty-eight of whom you do not
mention where, or by what persons they were vaccinated, if they ever have at all !   Besides, there are
twelve or thirteen individuals, supposed to have got
the cow-pox by *milking*, but no evidence to prove
that assertion ; and in such cases, who can be judges
as to the facts ?

Again, it is universally acknowledged by vaccinators, that if the small-pox actually infects a person,
prior to the time of his being vaccinated ; that is, if he
be under the influence of variolous contagion, before
the vaccine disorder takes effect in his constitution ;
no reliance can, in such cases, be placed in the
preventive power of vaccine inoculation : conse-
quently, no trials of vaccine inoculation, under such
circumstances, can be regarded as fair or valid.   Ne-

vertheless, I find not fewer than one hundred and
forty-five of your cases of failure to be apparently of
this kind! Some of these, you say, had both the small-
pox and cow-pox together; others did not escape
the ordinary progress of the variolous disorder;
and others seem to have been vaccinated prior to
their catching the small-pox, but you do not tell us
how long, or it was only a few days; and numbers of
the patients were inoculated for the cow-pox two or
three times, by your own confession, *without effect :*
so that in all these ONE HUNDRED AND FORTY-FIVE
INSTANCES, we cannot discover if the individuals had
really the *constitutional and local symptoms of cow-
pox previous to their exposure to variolous contagion!*
But we must learn to discriminate carefully between
mere *inoculation,* and actual *vaccination,* which you,
Sir, are perpetually confounding; since it will often
be requisite to inoculate repeatedly, before we are
assured that our patients have received the infection,
whether of small-pox or cow-pox.

This leads me to notice your disingenuous and
misplaced subtilty in eluding all inquiries, respecting the actual fact of vaccination, where the small-
pox has succeeded to an attempt at communicating
the vaccine disease. A simple endeavour to vaccinate,
may fail of vaccinating; and then we cannot hope for
any security whatever against the ravages of a loathsome infection by the small-pox, at some future
period! You, Dr. Bragwell, knew this, when you
thus required us to limit our questions :—" Have the

" parties been inoculated for the cow-pox ? *Yes.*—
" Have they had the small-pox afterwards ? *Yes.*—
" As to *How, When, Where,* Whether the Cow-
" pox *took,* was *genuine,* or *spurious* ; they are eva-
" sive and irrelative." You say, these inquiries
" may confound fools ;" and I may add, that none
but FOOLS of the most dangerous stamp would
endeavour to decide in any case, without making
such necessary inquiries. You, Sir, with no small
arrogance, pretend to chastise vaccinators who ask
those questions; and who (I think wisely) address the
" reporters with *How, What, When ?* Are you sure
" they went through a regular and genuine Cow-
" pox ?"—For, if such cautious investigators as
these have appeared to " *merit animadversion,*" it
is no wonder how smoothly and easily the number
of your " failures" have been accumulated to " five
" hundred and four proofs." But, *even this* is not
all I have to say against your boasted " SELF-EVI-
" DENT FACTS."

DOCTOR.—" May Heaven inspire other prac-
" titioners to follow my example!"

SURGEON.—Your example, Sir, has too often
been followed; and it is only by doing so that the
public mind has been, or can be agitated, and grossly
misled respecting vaccination.

DOCTOR.—" These, and similar delusions, can
" no longer avail ! The repeated attempts at de-
" ception on these grounds are no longer tenable ;
" the whole *reasonable* Faculty laugh and shake their

" heads at those daring violations of truth, and the
" miserable shifts to which vaccination has been re-
" duced."

SURGEON.—This insolent language will not
do, with those who know in what manner you begin
and finish your curious calculations, or *demonstrations*
against Cow-pox ; and I perceive (by that shake of
the head) what feelings the Rev. Gentleman has, on
hearing this exposure of your conduct. But, I have
not done with you, Dr. Bragwell ; although I have
pretty much reduced your catalogue of " five hundred
" and four PROOFS." What I next shall mention,
will shew that " vaccination cunning" is by no means
equal to the subtlety of " anti-vaccinarians."

A large proportion of your cases appear to be described *two or three times over*, in different parts of
your book, with slightly varying circumstances, as to
the address of patients, or the mode of spelling their
names, &c. ! ! ! It is not easy to detect and expose
every instance of this sort ; and possibly they are more
numerous than I am prepared to conjecture, from
the similarity and coincidence of your several detailed
accounts. This, however, is a point which can only
be elucidated by a perusal of your own statements ;
of which a few, to be hereafter mentioned,will establish
the charge itself, leaving it for your own conscience
to vindicate the other suspected cases ! The Rev.
Gentleman, perhaps, will take the trouble of examining all " this mass of evidence," in order to sa-

tisfy himself of these palpable errors, imperfections, falsehoods, and inconsistencies.

PARENT.—Enough has been already advanced to convict him of gross deception ! I shall take no further trouble in such a dirty affair !

SURGEON.—Allow me then to point out a few specific examples, by way of illustration and demonstrative proof. While you read the cases I shall refer to, it will at the same time be manifest how bald, and wretchedly incomplete these statements all are : and this should not be overlooked, in perusing the whole five hundred and four cases, as they are detailed by the Doctor. I address myself to you, Reverend Sir, as witness in this matter ; for it is not to be imagined that a culprit will plead guilty, even if his conscience admit the aggravated charge.

DOCTOR.—" *Humanum est errare.*"

SURGEON.—It is something to confess that you can *possibly* err : but whoever will glance at your original account, must perceive, Dr. Bragwell, that you were *conscious* of repeating the same stories two or three times over ; and that, in the crowd of histories you thought it fair play to multiply and subdivide them, so as to make up your sum total to " FIVE HUNDRED AND FOUR FACTS," or, as you express it on the title-page, " ABOVE FIVE HUNDRED " PROOFS OF FAILURE." I should like to hear how many proofs you would venture to insist on, after the deduction of these various sorts of exceptionable cases ?

5

I still have a word or two to offer on some of the remaining " *proofs of failure*," which you have triumphantly published.

DOCTOR.—Say on, Sir, say on. " Any errors
" in the statement of facts will be readily corrected;
" unless sophistical cavilling, subterfuge, prevarica-
" tion, or fallacy, be advanced, as usual. Neither
" cavilling, ribaldry, nor tart invective, have any
" effect on sensible minds searching for *truth*, espe-
" cially from men who seem to have banished de-
" cency and probity from the human virtues."

SURGEON.—These immoral qualities are not to be charged to the account of every person who (with great pains-taking) discovers your falsehoods, or " errors in the statement of facts :" nor is it to be expected that readers in general should have opportunity, ability, or inclination to search out and expose your *untruths :* but the cloven foot shews itself very manifestly at length! although not until you have deceived vast numbers of those whom you denominate " THE GAPING MULTITUDE."

Among the " errors" which may " be readily cor-
" rected," you will allow me to enumerate, in the next place, all the cases of cow-pox *after* small-pox, —cow-pox *pustular* eruptions,—*contagious* cow-pox, —children *never* inoculated,—some persons who have only *existed* in your own brain,—others who you admit " *did not take*" the vaccine disease,—some cases perfectly unintelligible,—and lastly, patients " who *now* come to the Mary-le-bone Infirmary,"

I

although reported to " HAVE DIED ! !" My remarks will be sufficiently exemplified, by a perusal and comparison of the following Cases : viz.

Bridges, 221, 375, 455.	Warren and Warrener, 180,
Broughton, 493, 304.	181, and 198.
Baker, 361, 374.	Bambridge, 28, 29, 30.
Cozens, 32, 457.	Collins, 175, 393.
Marshall, 132, 232.	Miles, 74, 328.
Macpherson, 222, 419.	Sadler, 288, 421.
Morgan, 257, 437.	Street, 139, 287.
Perch, 151, 360.	Willis, 441, 442.

Also 334, 448, 75, 481, 37, 38, 126, 112, 148, 357, 409, 390, 391, 482, 278, 279, 280, 281, 282, 283, 284, 79, 153, 154, 155, 156, 198, 200, 326, 36, 184, 214, 390, and 474.

All these cases should be read in the order I have enumerated them.

DOCTOR.—Perhaps, Gentlemen, I may not be able, after this *éclaircissement*, " to demonstrably " prove ALL the allegations in my book TRUE :" but, " knowing the cavilling character of the furious " vaccinators, I introduced many of the horrid cases " that have arisen from Cow-pox, at my two first " introductory lectures, the beginning of Octo- " ber 1805 ; and convinced every one present of " Cow-pox enormities."

SURGEON.—" Convinced every one present ?" Did there happen to be any person at your lectures who had inquired into the grounds and authorities upon which your allegations stand ? Was it known that most of them had been again and again refuted in print ? If not, you might easily carry on the quackish imposition, without fear of being detected by your group of uninformed spectators.

DOCTOR.—" The scene was truly affecting ! " *A load of children, brought in a cart from*"———

PARENT.—Indeed, Doctor Bragwell ! This was a very droll and singular invention, an admirable thought to attract notice from " the gaping multi- " tude." But, it might have done better at Smith-field, on St. Bartholomew's day.

DOCTOR.—" The first case brought into the " lecture-room, was Case 36; JOWLES, *the cow-* " *poxed, ox-faced boy,* who likewise has a terribly " diseased elbow-joint."

SURGEON.—I really must be excused staying any longer to hear this ludicrous story of a medical *puppet-show,* equalled only by the *stage-pranks* of Drs. Bossey and Yeldhall at Tower-hill or Moor-fields ! But, since you have introduced that won-derful, outlandish, " *beastly*" creature, with an hi-deous " OX-FACE," &c. and have been pleased also to publish an history (Case 390) of another child who had " *all the symptoms of the cow-poxed, ox-faced* " *boy, florid and frightful* ;" I shall request leave to make a short comment upon the text. And I may do this with the greater confidence, when it is men-tioned that the *former* " ox-faced" patient has been above four months past under my care, and that the *latter* case was inoculated by me for the cow-pox. Both justice and humanity, therefore, demand that I should be heard candidly on this occasion.

I suppose, Reverend Sir, you will allow that I have the use of my visual organs, as perfectly as the sharp-sighted Doctor: and yet I solemnly declare, that during the long time I have had an opportunity of inspecting these two " frightful" animals, they have appeared to possess the genuine characters of real human beings! Neither of them has had the smallest affinity or resemblance whatever to an *Ox*, any more than the learned *Ox*onian himself resembles an *Ass!* The curious picture, " so well and so " faithfully executed by the ingenious Mr. Pugh and " Mr. Annis," although admired " by all" the Doctor's " hundred auditors," conveys not the smallest likeness, to my imagination, of the bovine tribe : but it was a lucky and a bright thought, perfectly accordant to an original idea of Dr. MOSELEY, that the vaccinators were " transforming" the features and animal propensities of the human species! Dr. Moseley has, indeed, carried the opinion so far as even to hint at the possibility of the fair sex becoming at last enamoured of the *Bull*; and perhaps it might be in aid of this profound suggestion, that Dr. Bragwell has detected so ·many " *Ox*-faces," transformed by the " beastly *Cow*-pox." Truly, JOHN BULL will have great cause to bellow with rage after this discovery, and to be very jealous of his cornuted rivals!

Two days ago, when I examined the countenances of both these calumniated children, (Priest, the girl, and Jowles, the boy,) their " visages," most happily, were neither " terrible" nor " resembling an ox ;" so that

I trust our affrighted matrons will, hereafter, not
consider this *luminous* " FACT" of Dr. Bragwell as
either " visible or self-evident," nor half so credible
as the IONIAN TALE.

But the facility with which some weak people
may be induced to adopt the most absurd and ridiculous notions, propagated by the enemies of vaccination, is apparent from an anecdote recorded by
Mr. Ring, in his Treatise on the Cow-pox: namely, *of
a lady who complained to Mr. Simpson, that since her
daughter was inoculated, she coughs* LIKE A COW, *and
has grown hairy all over her body* ; *and Mr. Blair was
told, on a late excursion into the country, that the
inoculation of the cow-pox was discontinued there, because those who had been inoculated in that manner,
bellowed* LIKE BULLS !

Dr. THORNTON, in his *Vaccinæ Vindicia*, just
published, gives a detailed account of the " ox-faced
" boy," whose *real* disease (viz. *scrofula)* I have
described, and communicated in a letter to him. I
have also sent a slighter description of the same case,
to Dr. ADAMS and to Mr. RING ; who (as well as
Dr. T.) are sincere friends of mankind, and have
written works on the cow-pox which are too much
known to need my poor commendation. In particular, I ought to pay a tribute of respect, with the
rest of my countrymen, to the name of Mr. RING ;
whose various lively publications and fugitive pieces
form a concise history of vaccination, and anti-vaccinarian tricks, in all parts of the world. Nor ought I

(as the opportunity now occurs) to omit the names
of some other British authors, who have laboured in
the same field; and to whom I refer you, Reverend
Sir, for a more ample view of the Jennerian Controversy: *viz.* Dr. G. PEARSON, Dr. WOODVILLE, Dr.
LETTSOM, Dr. WILLAN, Dr. BARRY, Mr. H. JEN-
NER, (nephew of the DISCOVERER, whose own works
should first of all be read,) Mr. C. AIKIN, Mr.
PAYTHERUS, Mr. ADDINGTON, Mr. CREASER, Mr.
DUNNING, Mr. MOORE, Mr. MERRIMAN, Mr.
BRYCE, Mr. G. BELL, Mr. SHOOLBREAD, ACULEUS,
the Rev. Mr. WARREN; and, in a few days will be
added the name of another benevolently zealous
Clergyman, the Reverend ROWLAND HILL, whose
pamphlet (which I have seen in manuscript) is admi-
rably adapted for popular use, and whose efforts have
been peculiarly encouraging in the cause of " MILD
" HUMANITY, REASON, RELIGION, AND TRUTH."

By perusing their works, or only a fourth part of
them, you will discover the height, the length, the
depth, and the breadth, of *anti-vaccinarian frauds
and mistakes :* you will see, to your great astonish-
ment, " the mass of evidence" collected by Dr.
Bragwell in its truly disgusting deformity : you will
find, that most of the adverse " facts" which he and
others have been so actively holding up to public
notice, are repeatedly and " demonstrably proved"
to be either forgeries, or the well-meant fables of
incompetent judges : you will perceive how ingeni-
ously, perversely, and degradingly, these adventurers

have raked the very kennels of human misery, for
diseases; that they might impute them all to the
" beastly origin" of vaccination, and argue their
unenlightened readers into a persuasion of their own
disinterested exertions! It would exhaust your patience, and fatigue your attention, if I were to enter
into a description of the cases which I myself and my
friends have known to be thus egregiously misrepresented: I shall therefore decline, at present, de-
taining you with any farther observations of this kind.
Only let it be remembered, that I do not wish you
to take my *ipse dixit* in behalf of vaccination, but to
sift and examine the *whole* affair for yourself, as a
question of inconceivable importance to society.

My own experience can never stand in compe-
tition with that of many others, in this department of
medicine; but, such has been my success (in about
seven hundred cases), that it would be nothing short
of insanity to discontinue the practice of Jenner, and
to substitute the curse of variolous pestilence in its
stead: I should regard such conduct as no less criminal than voluntarily throwing fire-brands, arrows,
and death, among my peaceful fellow-creatures, in
the centre of this vast metropolis! Dr. Bragwell's
words, prophetically uttered on another occasion,
must now recoil upon himself—" *Calumny and de-*
" *traction,*" said he, " *have been, and will be, ex-*
" *ercised against every attempt to improve medicine.*
" *The more* SUCCESSFUL *any improvement is, with*
" *so much the more fury it is opposed; and it rarely*

4

" *fails to excite, in envious minds, private opposition,*
" AT THE EXPENSE OF HONOUR, INTEGRITY, AND
" TRUTH." Too exactly, alas! have these just observations been verified! For, if any improvement
be " SUCCESSFUL," it is this; which, under God,
hath stopped the horrid ravages of a fatal epidemic, in
Villages, in Towns, and in Cities; which hath arrested the footsteps of death, in its most horrid and
ghastly forms; which already (notwithstanding the
malevolence of some, and the culpable ignorance of
others) hath preserved its millions to the present ge-
neration, and will extend its benefits to posterity in
a tenfold proportion! Surely, in such a cause, the
name of JENNER will resound (and it even now
resounds through all this gloomy opposition) to the
farthest limits of the globe! If the fame of the
Cæsars and Alexanders, who have destroyed mankind by thousands, be transmitted to their children's
children; such a discovery as this will indeed com-
memorate the name of its AUTHOR (mild, modest, and
unassuming as he is) to the latest period of time! Oh,
that I had the tongue of Angels to proclaim his
praise!

DOCTOR.—" O tempora! O mores!"

PARENT.—*Ah! sly Deceiver! branded o'er and o'er,*
     Yet still believ'd !——

SURGEON.—I remember to have read, in Dr.
Woodville's excellent History of Inoculation, a remark
of this kind: that the writers against small-pox inoculation, when it was originally introduced into Eng-

land, instead of waiting to ascertain such facts as
might have enabled them to form just conclusions on
the advantages or disadvantages of this *new art*, immediately proceeded to employ falsehood and invective ; reproaching the inoculators with the epithets
of POISONERS and MURDERERS. Exactly so, have the
present worthy and impartial opposers of vaccination
acted ! One man (I knew him personally, and am
sorry to say he was a *Surgeon)*, whose income proved
much too narrow for his extravagance, used to fabri-
cate cases, and propagate the idle tales of every igno-
ramus, in the different newspapers, &c. His puffs
and egotisms were of the same complexion as those
of Dr. Bragwell ; and most of his " FACTS," although
" SELF-EVIDENT" to himself, were totally devoid of
evidence to persons possessed of common discernment. The unfortunate adventurer, not finding his
scheme sufficiently productive, and being unable to
preserve his reputation, put an end to his own mad
career, by an expedient too common among such
desperadoes ! This unhappy genius was the then
great champion of anti-vaccination, whose delusive
stories alarmed the public, and *first* gave occasion to
circulate a testimonial in favour of the Jennerian practice, July 1800 ; to which I had the honour of adding
my signature, among many others of superior worth
and notoriety. Next to this upstart, whose name I
would bury in oblivion, appeared Dr. Moseley, Dr.
Bragwell, and Mr. Birch : each of whom, prema-
turely and without any experience, endeavoured to

K

decry the merits of vaccine inoculation, before a Committee of the House of Commons ; when a host of *well-informed men* (as Dr. Moseley acknowledges), chiefly physicians and surgeons, *who had not made up their minds hastily, on a subject which so greatly concerned their characters and the dearest interests of society*, gave their opinions before that august assembly.   The result of this serious and momentous inquiry was, as all the world knows, THAT VACCINATION APPEARED TO BE A PROPOSAL OF THE UTMOST UTILITY,—THAT IT WAS A MILD SUBSTITUTE FOR THE SMALL-POX,—THAT IT IS INCAPABLE OF BEING COMMUNICATED BY CONTAGION,—THAT IT NEVER PRODUCES ANY OTHER DISEASE,—THAT IT NEVER OF ITSELF PROVES FATAL,—THAT IT MAY BE SAFELY PERFORMED, AT ALL TIMES, AND UNDER ALL CIRCUMSTANCES,—THAT IT TENDS TO ERADICATE THE SMALL-POX,—AND, IF UNIVERSALLY ADOPTED, MUST ULTIMATELY EXTINGUISH THE MOST DESTRUCTIVE DISORDER THAT EVER AFFLICTED THE HUMAN RACE ! ! !

PARENT.—Yes, Sir, I heard of all this, and that Dr. Jenner received, in consequence of such favourable and unequivocal evidence, the grateful, though inadequate remuneration of ten-thousand pounds, for his national discovery ; concerning which the late magnanimous statesman, Mr. PITT, observed, in the House of Commons, *that it was one of the most important ever made since the creation of man,— that the value of this discovery was, incontestably, with-*

out example, and beyond all calculation ! THE PRINCE
OF WALES, likewise, (not to speak of many other
Royal and Noble personages,) has borne his positive
testimony to its success and importance; *being tho-*
roughly persuaded, said His Royal Highness, *of its*
efficacy, and of the incalculable advantages the world
in general will reap from the indefatigable and praise-
worthy perseverance, with which Dr. Jenner has
brought it to its present perfection.

SURGEON.—It is well known, and deserving of
mature consideration, that no practitioners are (in
general) so difficult to satisfy, with the evidence adduced in favour of the cow-pox preventive, as those
who have been accustomed to partake of the profits
arising from small-pox infection aud inoculation. I
learn that the chief enemies of this new practice in
the East Indies (for it is introduced all over the
world) are the avaricious BRAMINS ; although the opposition of so powerful a set of persons has not been
able to prevent NINE HUNDRED THOUSAND
of the poor natives, in the British provinces of India,
from thankfully partaking of so inestimable a benefit !
But, there is nevertheless an exception to this general
selfishness of variolating Practitioners which ought.
not to be forgotten : I mean, in the liberal and humane conduct of Dr. Woodville, late Physician to the
Small-pox Hospital; who, in his answers to the
Committee of the House of Commons, declared that
he had encouraged vaccination at the Hospital, and
greatly preferred it to the small-pox—as it afforded

equal security against the future infection of that loathsome disease, without endangering life, or spreading contagion; and, that he had *already vaccinated* 7500 *persons*, half of whom were afterwards inoculated for the small-pox, *without* ONE OF THEM *taking this disorder ! ! !*

PARENT.—A pleasing and generous concession! It must be a complete reply to the groundless charge of " *interestedness*" in those who first promoted vaccination, which Dr. Bragwell so illiberally maintains; for certainly, a physician under such circumstances would expect to lose a large share of the emoluments of his profession.   This was not acting as if he hoped afterwards to " *revel in wealth*."

SURGEON.—The result turned out as might be expected. Dr. Woodville had long before (it is alleged on good authority) obtained about 1000*l.* per annum, by his profession : but, when the cow-pox inoculation was introduced, his yearly income sunk even to less than 100*l.*   Notwithstanding he was so great a loser, it is related of him that he paid the highest possible compliments and deference to Dr. Jenner, only a few days before his decease; and that he died pouring out blessings on this benevolent discoverer, whose new practice, he believed, would eventually save millions of lives !

If the rapid decrease of Dr. Woodville's income, and that of other active variolators (who now discontinue the practice), be considered, we have no reason to wonder at Dr. Moseley's declaration; that a certain *Hospital Surgeon was not willing to give up the practice of inoculation for small-pox.* No, no! This mine is too rich to be speedily neglected! The AURI SACRA FAMES predominates! It is not every man who can willingly " give up" a branch of practice which, like the boasted philosopher's stone, converts so much human calamity into GOLD! *Disinterested* philanthropy is not an ordinary virtue.

PARENT.—Nevertheless, you must not under-rate the philanthropy of medical men; which, on this occasion at least, has shewn itself to be eminently conspicuous, though by no means universal.

DOCTOR.—" THOSE," I tell you, " WILL BE " CONSIDERED THE GREATEST ENEMIES TO SOCIETY, " WHO LONGEST PERSIST IN SPREADING THE CRI- " MINAL AND MURDEROUS EVIL."

PARENT.—Undoubtedly they will. The only question is, Which has in fact proved itself to be the most " *criminal and murderous evil,*" the vaccine or the variolous disease?

SURGEON.—It would be loss of time, to demonstrate that the small-pox has destroyed its millions in every quarter of the globe: even the scanty portion of earth on which we Islanders breathe, loses about *forty thousand* of its inhabitants *every year*, by that terrible scourge! and in London alone, it has been found by melancholy experience, that near three thousand have died of it annually. Nothing but the late perverse, and too successful, attempts to spread

5

this direful pestilence, could have prevented a very great diminution of so " MURDEROUS AN EVIL."

DOCTOR.—I am for admitting only demonstrative facts. Can you " demonstrably prove" these assertions ?

SURGEON.—We argue, Sir, from analogy. *You* unreasonably keep up the flaming plague which *we* are endeavouring to extinguish ; but in other cities, towns, villages, &c. upon the Continent, the small-pox has, *in fact*, been nearly quite annihilated, or mitigated, beyond controversy. In England we cannot expect this, while you continue to disseminate the seeds of death with impunity. Would to God, the Parliament of England may speedily restrain, or at least regulate, such criminal proceedings, and crown our endeavours with complete success !

There are some large cities in which this beneficial change has been effected so gradually and progressively, as to afford the most pleasing prospect of a total eradication ; and of this I will give you one example in the city of Vienna, where the small-pox has prevailed dreadfully for many centuries. You will, from this decisive experiment, judge what might be done by the general practice of cow-pox inoculation in our *insular* kingdom, where so many thousand human beings have been annually cut off by the small-pox !!! In the year 1800, the total number of persons who died at Vienna was fourteen thousand six hundred, *of whom eight hundred and thirty-five were destroyed of the small-pox* ; but on introducing vaccination,

the number of deaths by the small-pox was immediately lessened; so that in the year 1804, *only two
persons died of that pestilential disorder!* We may
fairly call this an extinction of the small-pox; because of those two individuals who died, one was the
child of a boatman who caught the small-pox upon
the river Danube, and the other was a child sent from
a distant part of the empire to Vienna previously
infected. The gradual and beneficial effect of the
cow-pox may be observed by the following comparative statement of the deaths in that city, during
the period of five years:

A. D.	Total Deaths.	By Small-Pox.
1800	- 14,600	- 835
1801	- 15,181	- 164
1802	- 14,522	- 61
1803	- 14,583	- 27
1804	- 14,035	- 2

PARENT.—How manifestly beneficial! Nothing
can be more conclusive! This single experiment,
attended with such complete success, is more convincing than ten thousand opposing arguments and
negative facts! But why are we BRITONS, among
whom this great discovery originated, still so hor-
ribly scourged with variolous contagion, that (according to the printed bills of mortality) NO LESS
THAN NINE HUNDRED AND FIFTY DEATHS FROM THE
SMALL-POX OCCURRED IN LONDON DURING THE
LAST THREE MONTHS, OF 1805? And I am in-

formed that a similar fatality has been experienced in several other parts of the United Kingdom.

SURGEON.—This melancholy truth is easily explained. The last quarterly Report of our Royal Jennerian Society, (read Dec. 4, 1805,) has hit upon the true physical cause of so deplorable a cala-mity ; viz. THE CONTAGION OF SMALL-POX, DISSE-MINATED BY MEANS OF THE RENEWED AND GREATLY INCREASED PRACTICE OF INOCULATION FOR THIS DREADFUL DISEASE. It may clearly be proved, from the yearly bills of mortality in this metropolis, that a *greater number of persons have annually died in London of the small-pox* SINCE *the use of inoculation than* BEFORE ! The calculation has been taken on a large scale, for the period of eighty or a hundred successive years ; from which it is evident, on an average, that there are at least seventeen (some reckon even so many as twenty-one) deaths from small-pox in every thousand who die, *more than there were before* the practice of variolous inoculation ! So that if we allow eighty-nine deaths by small-pox, out of every thousand who have died in London, since the practice has been introduced ; there was a much smaller proportion than this, not less than seventeen in a thousand, before the use of inoculation : and so lamentable a fact can be explained only by admitting, that the partial inoculation of small-pox,(for it never was universal) keeps up a perpetual source of conta-gion in the air, which infects those who might other-wise have escaped the disease through life. The

casual small-pox is, without doubt, incomparably more
fatal than the inoculated; there being about *one in
two hundred* who die of the latter, and nearly the
proportion of *one in six* of the former: but it is nevertheless true, that within these few years the mortality from 'small-pox, ON THE WHOLE, has been
augmenting in London; for during the first thirty
years of the eighteenth century, in every thousand
deaths, the number of those destroyed by small-pox
was *seventy-four*; while in the last thirty years of
that century, the deaths from the same disease
amounted to *ninety-five* in every thousand!

DOCTOR.—This circumstance only shews that
small-pox inoculation ought to be universal, without
excepting one individual in the whole world. " Out
" of many thousands, nay, MILLIONS, it has been
" fully proved that scarcely ONE dies from small-pox
" inoculation."

SURGEON.—The universal adoption of small-
pox inoculation will not be effected until you can
persuade all men, at all times, to be of one mind;
and your assertion, respecting the safety of inoculated
small-pox, " *scarcely one dying in many millions*,"
is contradicted by all classes of persons who are capable of forming an opinion on facts which daily
occur: you might as well affirm that men are not
mortal, or that the sun has never shone at noon-
day.

DOCTOR.—" Salutary small-pox inoculation!
" To ascertain whether known, certain, and long-

L

" experienced small-pox inoculation, should be re-
" linquished for novel, doubtful, disastrous, and de-
" structive vaccination, is the object of my Essay.
" Let the Faculty, and the whole world, judge."

SURGEON.—Allow me to proceed a little, without interruption.   Your sentiments are too extrava-
gant to be received.

DOCTOR—" Unknown, greasy, horse-heeled
" practice, in opposition to small-pox inoculation ;
" which has been so improved and refined, as to be
" scarcely considered a disease !   Vaccination stands
" condemned by the experience of veterans in the
" profession."

SURGEON.— " Veterans" in error, Sir, are much
more dangerous to society than young men ; because
they are usually more pertinacious, and incomparably
less open to conviction.

DOCTOR.—" There requires a change of system
" in small-pox inoculating hospitals."

SURGEON.—I perfectly agree with you in this
observation ; though perhaps you would not carry
" the change of system" so far as I deem requisite :
for I am clearly of opinion, that the continuation of
" SMALL-POX INOCULATING HOSPITALS" is a fertile
source of misery and death in this vast metropolis.
I wish them to be set apart entirely and exclusively
AS PEST-HOUSES, to receive all those patients
who have casually become infected with small-pox ;
but never to be employed as " inoculating" establishments.   While the contrary practice obtains,

and is encouraged, we never can possibly exclude the small-pox from London ; because there always will be sordid and avaricious persons, interested in renewing the stock of contagion.

DOCTOR.—" Cow-poxing was introduced to
" prevent and exterminate the small-pox ! It has
" been a striking and uniform conduct in vaccina-
" tion, to augment the disasters of small-pox ;"
whereas, " *there is little danger in natural small-pox,*
" *if skilfully treated, and small-pox inoculation is as*
" *little alarming as vaccination !* The comparative
" view of cow-pox and small-pox inoculation, was
" read by me before the Honourable Committee of
" the House of Commons, deputed to examine cow-
" pox inoculation ; but my paper never appeared in
" the Report : what I did read and say, is for the
" most part suppressed ; and what it was impossible
" for me to say, has been published ! It appears
" then, that extermination is impossible, unless the
" vaccinators have more power than the ALMIGHTY
" GOD HIMSELF ;" and the very " supposition is
" IMPIOUS, UNTHINKING, IRRATIONAL, AND PRO-
" FANE! MY NEW TREATMENT IS SAFE, MILD, CER-
" TAIN, EXPERIENCED, SUCCESSFUL, AND UNAT-
" TENDED WITH DISASTER ! It seems that some of
" these busy vaccinators have lost their senses, their
" reason ; and are quite ignorant that the small-
" pox is contagious in the atmospheric air, inde-
" pendent of being communicated from one patient

" to another! The projects of these vaccinators
" seem to bid bold defiance to HEAVEN ITSELF, even
" to the WILL OF GOD.

" The order of the day is, to give surprising ac-
" counts from foreign doctors of small-pox extermi-
" nation. We need only say, *Look at home!* At
" the St. Mary-le-bone Infirmary, are two small-
" pox wards, for males and females : I have now
" under my care three young women, with the most
" malignant small-pox, and I am daily solicited for
" matter from my two wards. Nay, inoculation is
" gaining ground, even at the Small-pox Hospital!
" The Cow-pox inoculation has been persevered in
" for seven years ; but *no signs of extermination*
" *have appeared.* It has been affirmed, that the
" poorer people at Brighton drove their children
" last summer into houses where the natural small-
" pox was present ; and at another place, in Hamp-
" shire, the poor villagers were *inoculated for small-*
" *pox, by a farrier, at five shillings a head.*"

SURGEON.—While you and your agents re-
proach vaccinators for not accomplishing their humane intentions in England, as well as on the Continent, you strive in every possible way to prevent
their success, by disseminating the poison of small-
pox ! Is this conduct fair and honourable, Dr. Bragwell ?

DOCTOR.—" It is the duty of HONOURABLE
" MEN to alarm mankind ; who are returning to well-

" known, well-established, small-pox inoculation,
" WITH ARDOUR."

SURGEON.—The zeal of many " farriers" and
such-like " honourable men," in spreading a direful
pestilence, for the sake of a LITTLE PELF, is acknowledged and deeply lamented ! The insatiable love
of money unhappily prevails ! But if your suggestion
be true, that " EVEN AT THE SMALL-POX HOSPITAL"
this scourge of human nature continues to be pur-
posely inflicted, and sent forth to desolate the inhabitants of London, I know not how to reconcile such
proceedings with the original design of its Governors ! Much less can I reconcile such conduct with
the last printed Report of this benevolent Associ-
ation ; who (by their medical officers) vaccinated
above two thousand persons at the hospital during the
year 1805, and have since openly declared that ITS
CONSTANT SUCCESS HAD CONFIRMED
ALL THAT WAS PROMISED BY SO INVA-
LUABLE A DISCOVERY.

This public and unequivocal testimony, sanctioned by my friend Dr. Adams (who succeeded the
late humane vaccinator Dr. Woodville, and has also
written in defence of vaccine practice), induces me to
hope that nothing but the most imperious circumstances constrain the governors of that institution to
fan the dying embers of variolous contagion ! It is
nevertheless to be exceedingly deplored, that any
circumstance, however urgent and irresistible, should
authorise loathsome masses of living corruption to

3

be daily carried through the streets of London ! ! !
So horrid an exercise of British liberty, in spreading
a destructive plague at noon-day, and affording an
exhaustless fund of morbid poison, calls loudly *indeed*
for legislative interference and restriction ! ! !

But, the man who pretends to convert small-pox
into a " *mild*, and *safe*" disease, " *unattended with*
" *disaster*," is less entitled to belief, than the deluded
Alchymist, who promises to change the baser metals
into gold ! YOU, Dr. Bragwell, can have no excuse,
or shadow of an excuse, for your vain trumpeting ;
when the lives of thousands and millions, in every
clime, incessantly cut off by the variolous plague,
demonstrate the wilful falsehood of your cruel de-
clamation ! It is totally impossible that you should
attach credit to your own words ! YOUR CRIME
is not perpetrated in ignorance ; but must be the
fruit of old established habits, too bad and too inve-
terate to be eradicated ! The fact of small-pox being
a comparatively MODERN DISEASE, proves that the
idea you have adopted, respecting its production in
the atmosphere, is entirely chimerical : for otherwise,
the ancients (who have breathed the same kinds of
contaminated air as we) would have experienced the
same tremendous visitation.

How little your opinions have influenced those
who know you best, must be evident ; when it is
considered that, even among the Governors of an
extensive parochial infirmary, where you have many
years presided as physician, your medical character is

despised—your proffered advice is rejected—your opposition to the Cow-pox is lamented—and, as if to crush the hydra of malevolence, *peremptory orders have been lately given by your respected and intelligent fellow-parishioners of Mary-le-bone, to vaccinate all the children whom you would have wantonly exposed to variolous inoculation!*

DOCTOR.—" Let mankind arouse from their " lethargy, and chase from their houses all who pro- " pose vaccination :" I myself will be the foremost to set so laudable an example. Off, off, ye furious Cow-poxers ; this moment, be gone from my presence. Here's the door ; depart.

PARENT.—I now understand the true meaning of Dr. Bragwell's own words : " MILD HUMANITY, " REASON, RELIGION, AND TRUTH, MEET IN COM- " BAT AGAINST FIERCE, UNFEELING FEROCITY, " OVERBEARING INSOLENCE, MORTIFIED PRIDE, " FALSE FAITH, AND DESPERATION."

SURGEON.—Candid and Reverend Sir, you see what kind of a REASONER we have had to deal with! It now remains for you to act conformably with your present determination ; and to endeavour (by following the printed instructions for vaccination, which you may receive *gratis* at the Royal Jennerian Society) to rescue from the grave not only your own children, but those of your kinsmen and neighbours, who will hereafter look up to you as a wise counsellor and physician, as well as a spiritual father.

I ought not to part from you, without affirming

in the words of Dr. JENNER, from whom I this day
received a letter, that VACCINATION FLOU-
RISHES, notwithstanding the artful proceedings of
its enemies: it *flourishes* even in Great Britain,
where its foes have been more perseveringly bitter
in their opposition than elsewhere; it *flourishes* at
the numerous Institutions established for its support
in and near this Metropolis, where there have been
(so far as I can estimate them) at least one hundred
thousand persons vaccinated; it *flourishes* in every
European kingdom, and almost in every considerable
town upon the Continent, with astonishing success;
it *flourishes* in the remotest parts of the earth—in the
United States of America—in Mexico—in Canada—in
Peru—in Turkey—in several African provinces—and
throughout the vast dominions of India, both East
and West; in short, it *flourishes* wherever Britons have
had a local habitation or a name! THE MILLIONS
WHO HAVE BEEN ALREADY VACCINATED
WITH PERFECT SUCCESS, IN ALL QUAR-
TERS OF THE WORLD, AFFORD THERE-
FORE A CONVINCING PROOF, THAT THE
LESS FAVOURABLE EXPERIENCE OF A
FEW PRIVATE INDIVIDUALS AT HOME,
DEPENDS ON SOME PECULIAR CIRCUM-
STANCES, NOT JUSTLY IMPUTABLE TO
VACCINATION.

The only objection, which can, (with a shadow
of reason) be urged against the Cow-pox, is, that it
does not prove an INFALLIBLE security against the

small-pox, there being a FEW instances of small-pox occurring after vaccination. I will give this objection its full force, and argue upon the supposition of its being true : but, let me ask, what will be the consequence? Does the Cow-pox fail oftener than the occurrence of small-pox *twice in the same subject?* Probably these events happen in nearly a similar proportion of cases.

Again, suppose *fifty* persons have really had the small-pox subsequent to vaccination, out of three hundred thousand individuals vaccinated in this kingdom ; even thus, it will appear that only one person in six thousand is liable to have the small-pox afterwards : whereas, at least one person in three hundred (and above double that proportion in London) DIES of the inoculated small-pox ! But, in fact, not more than *ten* admissible cases of failure can be reckoned out of three hundred thousand individuals, properly vaccinated ; so that only one such person in thirty thousand is liable to the small-pox, and ten in three hundred thousand, the whole number supposed to have been already vaccinated in Great Britain.

Now, granting that of those ten, who catch the small-pox after vaccination, TWO should die, it would then follow that *only two individuals die out of three hundred thousand* persons, in consequence of failures in the inoculated Cow-pox ; whereas, the number of deaths from so many people inoculated for the small-pox would have been at least one in three hundred, that is, *a thousand in all!* Conse-.

M

quently, it is evident from this plain calculation, that
the deaths occasioned by small-pox inoculation, are
AT LEAST FIVE HUNDRED TO ONE MORE
THAN FROM THE COW-POX!!!

But, if we reckon the small-pox to happen twice
in the same person, as frequently as that disease
occurs after vaccination, this proportion will be
DOUBLED IN FAVOUR OF THE COW-POX;
and if we allow that variolous inoculation has been
the remote cause of the casual small-pox becoming
so universally prevalent and fatal as it now proves to
be, the advantages of vaccination (which proposes entirely to extinguish the contagion of small-pox) must
appear great indeed BEYOND ALL COMPUTA-
TION!!!

Allow me to say, in conclusion, that the issue of
THE VACCINE CONTEST is no longer doubtful;
and that the *comparative advantages* of Cow-pox
inoculation may be summed up in a few words:

I. THE INOCULATED COW-POX is so very
mild as not to deserve the name of a disease. It is
certainly not contagious; and, in the opinion of the
most experienced practitioners, has never once proved
fatal, when properly conducted.

II. It occasions only a pustule on the inoculated
part, and communicates no loathsome or hazardous
disorders: on the contrary, it has even been known
to improve health; and to remedy those constitutional complaints, under which the patients before
laboured.

III. It leaves behind it no personal blemish, but A BLESSING—one of the greatest bestowed on man, —*a perfect security* (in more than 9,999 Cases out of 10,000) *against the future reception of the Small-pox.*

I. THE INOCULATED SMALL-POX is loathsome, contagious, painful, hazardous, sometimes fatal; and, when partially adopted, spreads the variolous poison, which increases the general mortality.

II. It may occasion the same maladies as the natural small-pox; requiring also a course of medicine preparatory to inoculation, as well as during the disorder, and often for a long time afterwards.

III. It frequently leaves behind it the same blemishes, privations, and frightful deformities as the natural small-pox; which are the more deplorable, as they were brought on by a voluntary deed.

I. THE NATURAL SMALL-POX is a disgusting, contagious, painful, and very dangerous disease. It is confined to no climate; but rages in every country, and destroys nearly a tenth part of mankind.

II. Those who survive the ravages of this dreadful distemper, often live only to be the victims of scrofula, consumption, and other maladies; or to drag out a lingering existence, more terrible than death.

III. That merciless pestilential disorder, leaves behind it pits, scars, deafness, and loss of sight; with various other bodily deformities, which embitter life.

From this faithful statement of the advantages attending VACCINE INOCULATION, founded on MILLIONS OF CASES; it must appear manifest to unpre-

judiced persons, that it is the duty as well as **the**
interest of every parent, of every family, and of every
nation, immediately to adopt the new practice, and
thereby to hasten THE TOTAL EXTERMINA-
TION OF THE SMALL-POX.

Reverend Sir ; let me now entreat you to go home,
and address your parishioners in some such language
as the following :

HINTS TO PARENTS,

ON THE

COMPARATIVE ADVANTAGES

OF

SMALL-POX AND COW-POX
INOCULATION.

[*N. B. The Author has printed these* HINTS *as a separate Tract, and
at a very low price, for general distribution.*]

" You who are blessed with the gift of children, are no less bound
by the Christian religion, than prompted by parental affection, to
preserve your offspring from every threatening evil; and particularly
from those destructive pestilential diseases which may be caught, either
by infectious air, or the breath of contaminated persons.

" If a plague or a putrid fever raged in your neighbour's family,
you would deem it a crime to expose your children to its contagious
influence ; and, if you knew any mode of securing your babes from
infection, through life, how anxious would you be to use such means
as early as possible !   Then let me ask, if you ever considered the
small-pox as a dreadful plague, much more dangerous than a common
putrid fever ?   Do you not know, that the small-pox kills every year
about forty thousand inhabitants of this kingdom, little as it is in
comparison of many empires ?   Did you never learn that, in London
alone, nearly three thousand individuals are annually cut off by this
single disease ?   Surely it is impossible not to have heard, that the natural small-pox destroys a sixth part of those who catch it ; and
that considerably more than half the population of this Island have

the disease, during one period or another of their lives! Those who
recover from it, are too commonly disfigured by visible marks;
some have their constitutions ruined, or their limbs completely dis-
abled; while others suffer the loss of hearing or sight, and are ever
afterwards afflicted with the scrofula, or consumption, &c. &c. so as
to drag on a miserable and burthensome existence!

" Should you think to prevent all those distressing consequences
by inoculating your children for the small pox; it is to be remembered
that the same kinds of disorders may even then follow the inoculation, although the danger is thereby much lessened. However, there
is a distressing consequence which, perhaps, you may not always
think of, commonly produced by inoculation: Your neighbours
may, by that means, be exposed to a distemper which did not before
exist among them; the infant you inoculate may happen to escape
with its life, but the children of others may catch the small-pox from
your babe, and perish! Let me first notice the danger of your own
child's situation; since one in two hundred generally, and often a
still greater proportion, will die of the small-pox, when produced by
inoculation: and the circumstance of your having voluntarily exposed your child to a fatal disease, would be always a bitter subject
of reflection to your tender feelings. At best, recollect how much
nursing, and medicine, and precaution are required, to support an
infant under such a very afflicting and loathsome complaint as the
inoculated small-pox; and after all, you know there is some pro-
bability of your child dying, as it were by your own hands! But,
likewise, give me leave to impress your mind with the hazard you
subject your fellow-creatures to, by bringing an infectious disease
among them. The doing this without deliberation, is wantonly to
expose your neighbour's life to a certain and immediate danger; in
order to escape an absent evil, which God might, possibly, have kept
at a distance as long as your child lived!

" Suppose, for a moment, that you could prevent the small-pox from
infecting your child in the natural way *by paring its nails*, as certainly
as you could by inoculating it for the small-pox; would you not be
deemed insane, to refuse performing so slight an operation? Most
undoubtedly you would. And, if I could tell you of another little
operation, which only raises a small blister on the arm, and would
not occasion any illness in your child, but is quite as certain (as the
inoculated small-pox) of preventing all future infection; I am sure

you could not be so cruel to your infant, as to neglect this simple
mode of prevention. Such refusal on your part, would arise either
from ignorance or perverseness; and, after what I next will tell you,
it shall not be possible to plead ignorance.

" The *Cow*, from which your children and yourselves receive nou-
rishment, is sometimes troubled with an eruption upon her nipples,
called the COW-POX; and if, either through accident or design,
any person gets this disorder by imbibing a little of the matter dis-
charged from the eruption upon the cow, the small-pox can hardly
ever be communicated to that person so long as he lives ! This fact
has been proved by thousands and millions of trials: there is the same
security, if a particle of this matter be taken from one human being
so affected, and then inoculated by a scratch upon another person,
without coming directly from the teat of the cow; which indeed is
the method now practised almost all over the world, according to the
directions of *Dr. Jenner*, who has been rewarded with ten thousand
pounds (by the British Parliament) for so invaluable a discovery.

" The COW-POX or VACCINE INOCULATION (as this new
practice is named) causes no sickness or fever, of any consequence ; it
is never fatal, or even dangerous; it produces no other bad disorders;
it requires no preparatory medicines; it may be performed upon old
and young, or upon pregnant women, at all seasons of the year, with
perfect safety; and can only be given by inoculation, so that no one
can possibly catch it by breathing in the same room, or sleeping in the
same bed, with a vaccinated person: and therefore, it is not to be regarded in the light of a serious disease, but as an invaluable blessing !
The only appearance it makes, is that of a pock or vesicle on the
inoculated part, surrounded by a moderate degree of inflammation ;
which goes off gradually in a few days, and leaves first a little brown
scab, and then a small roundish scar like the inoculated small-pox.
By this easy and harmless mode, then, it is now in your own power
to prevent your children from catching that loathsome disorder,
which kills nearly a tenth part of mankind, without at the same time
hazarding the safety of your neighbours or friends !

" It must be admitted that the new inoculation has met with some
opposition; and objections have been urged, which it would be un-
candid to pass over in silence. Some of these objections, and certainly the most weighty, relate to matter of fact and actual observation; others have a more vague and doubtful ground. It is unques-

tioned, that some cases have occurred, in which a *careless* inoculation of cow-pox has failed to produce the promised security : other cases have been met with, in which the symptoms of the complaint induced by inoculation are stated to have been so severe, and even fatal, as to perplex those who had been accustomed to view in the new practice nothing but an uniformly mild, safe, and effectual preventive of a most formidable contagion. In answer to those objections, it might be urged, that were ALL the alleged instances of ill success acknowledged to be true in their fullest extent, and the mildness of cow-pox allowed to be only proportional; still this proportion, compared with the most favourable inoculation of smallpox, would give the new practice a decided claim to the preference of individuals, whilst its uncontagious nature (which is not disputed) would equally recommend it to public approbation.

" But it would be highly unjust to the merits of cow-pox inoculation, to make this allowance : for, the action of cow-pox cannot prevent the constitution from being at the same time attacked by infantile and other prevalent diseases ; so that the few cases of fatal termination imputed to this source, may fairly be ascribed to the concurrent operation of some mortal disorder, wholly unconnected with the new inoculation. It is the more candid to admit of this explanation, since by far the greater number of the supposed failures have been actually traced to some evident misrepresentation of facts ; or have been most satisfactorily accounted for, from the want of experience to ascertain the characteristic marks of the *vaccine pustule*.

" Two cases of death by the cow-pox having been published in the London bills of mortality, a committee of the Royal Jennerian Society was appointed to investigate their particulars. The committee reported, on the most authentic and satisfactory documents, that there was not the slightest foundation for attributing these instances of fatality either immediately or remotely to the cow-pox. One of the children died of scarlet fever, twelve months after it had passed through the vaccine disease ; the other, of convulsions from teething, fourteen weeks afterwards : in both, the inoculation was attended by none but the most favourable circumstances. These palpable errors arose with the *ignorant Searchers;* who (with no proper knowledge of the subject) have also reported a single death to have happened from the Cow-pox, during each of the years 1804 and 1805.

" To conduct with safety and propriety this inoculation, simple

S

as it is, an accurate knowledge of its genuine appearances, and of the spurious varieties which now and then exist, is indispensably necessary. Its simplicity has introduced some degree of carelessness in attending to its real character; some precautions, not at first noticed, are now found to be requisite: for, as the success of all medical practice has experience for its basis, it would have been wonderful indeed, and next to miraculous, if every circumstance relating to the vaccine inoculation had been at once suggested to the minds of its earliest promoters, by intuitive discernment.

" Certain opponents of the new practice have spread an alarm of some transformation, some mysterious change in the very nature and propensities of the human race, to be apprehended from the introduction of a disease originating in a brute animal! To such an absurd idea as this, which has never been fairly brought forward, and indeed seems almost entirely abandoned; I need only say, in reply, that it is unsupported by a single fact, or probable analogy, and is actually destroyed by the experience of time immemorial, in the countries where cow-pox was first discovered. Since then, the infection derived immediately from the Cow is found completely free from these objections, and since successive inoculations from one human subject to another have hitherto produced no other effect than to mitigate all the symptoms that attend the original disease; how can it be thought presumptuous or rash, to root out from amongst us a pestilential and destructive malady, by the substitution of a mild and benign disorder, taken from an animal entirely devoted to the service of mankind?

" After these remarks, I hope you will conscientiously use so providential a method of rescuing your offspring, and neighbours, from the small-pox contagion; which, for many centuries, has dreadfully alarmed and afflicted the whole human race."

THE END.

ERRATUM.
Page 35, line 23 ; for *seventy-two*, read *one hundred and eighty*.

S. GOSNELL, Printer, Little Queen Street.

Printed in the United States
By Bookmasters